GENERAL ENGINEERING SCIENCE FOR TECHNICIANS

THIRD EDITION

SI UNITS

General Engineering Science

FOR TECHNICIANS

THIRD EDITION
SI UNITS

R. J. BESANKO

F. I. Plant E.
Lecturer in the Department of Engineering, Brooklyn Technical College, Birmingham.
Chief Examiner in Engineering Science and Mathematics (Mechanical Engineering Technicians Course: Second Year) to the Union of Educational Institutions.

AND

T. H. JENKINS

B.Sc., A.M.B.I.M., A.F.I.M.A.
Vice-Principal, Foley College of Further Education, Stourbridge.
Chief Examiner in Engineering Mathematics (General Course in Engineering: Second Year) to the Yorkshire Council for Further Education.

OXFORD UNIVERSITY PRESS
1976

Oxford University Press, Ely House, London, W.1

GLASGOW NEW YORK TORONTO MELBOURNE WELLINGTON

CAPE TOWN IBADAN NAIROBI DAR ES SALAAM LUSAKA ADDIS ABABA

DELHI BOMBAY CALCUTTA MADRAS KARACHI

KUALA LUMPUR SINGAPORE HONG KONG TOKYO

ISBN 0 19 859143 8

© Oxford University Press, 1976

First edition 1968
Second edition 1970

Text set in 9/10 pt Monotype Times New Roman, printed by photolithography,
and bound in Great Britain at The Pitman Press, Bath

PREFACE TO THIRD EDITION

The Technician Education Council (TEC) was established in March 1973 in fulfilment of a recommendation of the Haslegrave Committee on Technician Courses and Examinations, which reported in 1969. When the creation of TEC was announced in the House of Commons by the then Secretary of State for Education and Science, its terms of reference were summarized as follows:

'The Council will be concerned in the development of policies for schemes of technical education for persons at all levels of technician occupations in industry and elsewhere. To this end it will, as proposed in the Haslegrave Report, plan, administer and keep under review the development of a unified national system of courses for such people and will devise or approve suitable courses, establish and assess standards of performance, and award certificates and diplomas as appropriate.'

This book is written especially for students taking TEC Certificate and Diploma Courses and covers the fundamentals of engineering science (mechanical and electrical). The book will also be useful in spheres other than colleges doing TEC courses. The basic subjects covered are Statics, Dynamics, Heat, Electricity, Magnetism, and Chemistry, and within this field the work will be of value for G.C.E. O-level subjects such as Science (Building and Engineering) and General Engineering Science in secondary schools.

Particular attention has been paid to the correct use of units, symbols and abbreviations. The system of units used in this edition is the Système International d'Unités (SI)—International System of Units—adopted by the General Conference of Weights and Measures and endorsed by the International Organization for Standardization. Certain non-SI units have been included where their use is permitted. The special name pascal (symbol Pa) has been approved for the SI unit of pressure and stress, N/m^2, and since this new name and symbol have only recently been adopted by CGPM it will take time for them to become well known and widely used. As a general principle the use of the pascal is to be encouraged and it is likely to replace the explicit expression newton per square metre (N/m^2) in many applications. In this edition the well-known N/mm^2 and the newly adopted MPa are used side by side for stress properties of materials. The pascal and approved non-SI units such as the millibar are used for pressures. Each chapter contains sufficient theory, description, and explanation to provide the readers with enough material to satisfy their requirements at this early stage, and to lay a firm foundation for further studies.

Numerous worked examples are included at frequent intervals throughout the book. Each of these examples has been carefully chosen, and the solutions presented in a detailed and easy-to-follow manner. At the end of each chapter there is a wide selection of graded problems most of which have been taken from past examination papers of the main examining bodies. An important feature is the use of a numbering system which correlates the numerous problems with the most relevant sections of the text; and this enables the student to locate the appropriate passage which he should read again if he is unable to answer a particular question. The numerical answers to worked and unworked problems are, in general, given to 'slide-rule accuracy'.

R. J. B.
T. H. J.

Birmingham
August 1975

ACKNOWLEDGEMENTS

We gratefully acknowledge the co-operation given by the following Examining Bodies in granting permission to include a selection of questions from past examination papers and for allowing us to modify some of these questions in order that the terms and units may be in agreement with the International System of Units (SI).

City and Guilds of London Institute (C.G.L.I.)
East Midland Educational Union (E.M.E.U.)
Northern Counties Technical Examinations Council (N.C.T.E.C.)
Union of Educational Institutions (U.E.I.)
Union of Lancashire and Cheshire Institutes (U.L.C.I.)
Welsh Joint Education Committee (W.J.E.C.)
Yorkshire Council for Further Education (Y.C.F.E.)

Our special thanks are also due to Crompton Parkinson Limited, Alkaline Batteries Limited, Mallory Batteries Limited and Electric Power Storage Limited, for kindly allowing us to include diagrams and details of their products.

Finally we express sincere appreciation to the staff of the Oxford University Press for their guidance and advice during the preparation of the book, and to our wives for their encouragement and for typing and checking the manuscript.

R. J .B.
T. H. J.

CONTENTS

1 Measurements and units

1.1. INTERNATIONAL SYSTEM OF UNITS

Measurement is the determination of size or amount in relation to a standard. There are standards, known as units, for all quantities such as distance, time, mass, and capacity.

The system of units of measurement used in this book is the Système International d'Unités (SI)—International System of Units—adopted by the General Conference of Weights and Measures and endorsed by the International Organization for Standardization. SI units are of three kinds: base, supplementary, and derived. The SI is based upon seven fundamental units, namely,

Quantity	Unit
length	metre (m)
mass	kilogram (kg)
time	second (s)
electric current	ampere (A)
thermodynamic temperature	kelvin (K)
luminous intensity	candela (cd)
amount of substance	mole (mol)

The two SI supplementary units are,

Quantity	Unit
plane angle	radian (rad)
solid angle	steradian (sr)

Approved definitions of the seven base units and two supplementary units are given below, but several of these definitions are beyond the scope of this book.

The *metre* is the length equal to 1 650 763·73 wavelengths in vacuum of the radiation corresponding to the transition between the levels $2p_{10}$ and $5d_5$ of the krypton-86 atom.

The *kilogram* is the unit of mass; it is equal to the mass of the international prototype of the kilogram.

The *second* is the duration of 9 192 631 770 periods of the radiation corresponding to the transition between the two hyperfine levels of the ground state of the caesium-133 atom.

The *ampere* is that constant current which, if maintained in two straight parallel conductors of infinite length, of negligible circular cross-section, and placed 1 metre apart in vacuum, would produce between these conductors a force equal to 2×10^{-7} newton per metre of length.

The *kelvin*, unit of thermodynamic temperature, is the fraction 1/273·16 of the thermodynamic temperature of the triple point of water.

The *candela* is the luminous intensity, in the perpendicular direction, of a surface of 1/600 000 square metre of a black body at the temperature of freezing platinum under a pressure of 101 325 newtons per square metre. (1 newton per square metre = 1 pascal).

The *mole* is the amount of substance of a system which contains as many elementary entities as there are atoms in 0·012 kilogram of carbon-12. When the mole is used, the elementary entities must be specified and may be atoms, molecules, ions, electrons, other particles, or specified groups of such particles.

The *radian* is the plane angle between two radii of a circle which cut off on the circumference an arc equal in length to the radius.

The *steradian* is the solid angle which, having its vertex in the centre of a sphere, cuts off an area of the surface of the sphere equal to that of a square with sides of length equal to the radius of the sphere.

The seven physical quantities, length, mass, time, electric current, thermodynamic temperature, luminous intensity, and amount of substance, are by convention regarded as being dimensionally independent. For any other quantity, the SI unit is derived by a dimensionally appropriate combination of the base units and supplementary units. For example, angular velocity equals angle turned through divided by time, so that the SI derived unit of angular velocity is the SI supplementary unit of plane angle divided by the SI base unit of time, i.e. radians per second (rad/s). Similarly, quantity of electricity (or electric charge) equals electric current multiplied by time, so that the SI derived unit of quantity of electricity is the SI base unit of electric current multiplied by the SI base unit of time, i.e. ampere-second (A s), which is given the special name of coulomb (C).

1.1.1. Multiples and submultiples of basic units.
For large and small quantities it is convenient to use multiples and submultiples of the basic units. Only one prefix is applied at a time to a given unit. For example, one thousandth of a milligram is referred to as 1 microgram (1 μg), not as 1 millimilligram (1 mmg). [See Table 1.1, page 2.]

1.2. SOME QUANTITIES AND THEIR UNITS

1.2.1. Length.
The measurement from end to end of a body or from point to point in space is called length or distance.

Distance is measured in units such as the metre (m) and the kilometre (km).

1.2.2. Mass.
The mass of a body is the quantity of matter in it. Matter occupies space and can be solid, liquid or gaseous.

The mass of a body never varies; it is constant under all conditions.

The kilogram (kg) and tonne (t) are two units of mass. 1 tonne = 1000 kg.

1.2.3. Time. The measurement of duration from one event to another is called time.

The second (s) is the basic unit of time, but certain non-SI units such as the minute (min) and the hour (h) will be used in many practical cases.

1.2.4. Weight. The force of gravity acting on a body is called the weight of the body. This force may vary depending on the position of the body. Weight, due to the force of gravity, is measured in units of force such as the newton (N) and kilonewton (kN).

The force of gravity acting on a body of mass m (kilograms) is equal to mg (newtons), where g is the acceleration due to gravity in metres per second squared (m/s²) (see Chapters 8 and 9).

1.2.5. Area. Area is defined as extent of surface, and is measured in square units of length—e.g. square metres (m²) or square millimetres (mm²).

1.2.6. Volume. The amount of space occupied by a body is known as its volume.

Volume is measured in cubic units of length such as cubic metres (m³) or cubic centimetres (cm³).

1.2.7. Capacity. Capacity may be defined as cubic content. It is generally considered as the amount of space available within a container, and is measured in litres (l).

The word 'litre' is used as a special name for the cubic decimetre. Thus 1 l (litre) = 1 dm³ (= 0·001 m³).

TABLE 1.1

The following table shows the prefixes denoting decimal multiples or submultiples.

Multiple or Submultiple	Prefix	Symbol
$1\ 000\ 000\ 000\ 000 = 10^{12}$	tera	T
$1\ 000\ 000\ 000 = 10^{9}$	giga	G
$1\ 000\ 000 = 10^{6}$	mega	M
$1000 = 10^{3}$	kilo	k
$100 = 10^{2}$	hecto	h
$10 = 10^{1}$	deca	da
$0·1 = 10^{-1}$	deci	d
$0·01 = 10^{-2}$	centi	c
$0·001 = 10^{-3}$	milli	m
$0·000\ 001 = 10^{-6}$	micro	μ
$0·000\ 000\ 001 = 10^{-9}$	nano	n
$0·000\ 000\ 000\ 001 = 10^{-12}$	pico	p
$0·000\ 000\ 000\ 000\ 001 = 10^{-15}$	femto	f
$0·000\ 000\ 000\ 000\ 000\ 001 = 10^{-18}$	atto	a

1.3. DENSITY AND RELATIVE DENSITY

1.3.1. Density. The density of a body is defined as the *mass per unit volume*, e.g. the number of kilograms per cubic metre (kg/m³) or grams per cubic centimetre (g/cm³).

$$\text{Density } (\rho) = \frac{\text{mass } (m)}{\text{volume } (V)}.$$

EXAMPLE 1.1. A piece of cadmium of mass 550 g is immersed in water in a measuring jar and found to have a volume of 64 cm³. Calculate the density of the cadmium.

SOLUTION

$$\text{Density} = \frac{\text{mass}}{\text{volume}},$$

$$\therefore \text{ density of cadmium} = \frac{550 \text{ g}}{64 \text{ cm}^3} = 8·59 \text{ g/cm}^3$$

$$= 8590 \text{ kg/m}^3.$$

1.3.2. Relative density. The relative density of a substance may be defined as the ratio:

$$\frac{\text{density of the substance}}{\text{density of pure water at 4 °C (277 K)}}$$

or

$$\frac{\text{mass of the substance}}{\text{mass of an equal volume of water}}.$$

It should be remembered that the density of pure water is a maximum at 4 °C, and at that temperature 1 gram occupies 1 cubic centimetre, so that the density of water is 1 g/cm³. It follows that the density of water is also 1000 kg/m³.

EXAMPLE 1.2. A bar of brass is 304 mm long and has a section of 25 mm × 6 mm. Determine the mass of the bar given that the relative density of brass is 8·48. Take the density of water as 1000 kg/m³.

SOLUTION

$$\text{Relative density } d = \frac{\text{density of brass}}{\text{density of water}},$$

$$\therefore \text{ density of brass} = \text{relative density of brass} \times \text{density of water}$$

$$= 8·48 \times 1000 \frac{\text{kg}}{\text{m}^3}$$

$$= 8480 \text{ kg/m}^3.$$

$$\text{But, density} = \frac{\text{mass}}{\text{volume}},$$

$$\therefore \text{ mass of brass} = \text{density} \times \text{volume},$$

$$\text{where volume of bar} = \text{length} \times \text{sectional area}$$

$$= 304 \times 25 \times 6 = 45\ 600 \text{ mm}^3$$

$$= \frac{45\ 600}{10^9} \text{ m}^3 \ (10^9 \text{ mm}^3 = 1 \text{ m}^3).$$

$$\text{Mass of bar} = 8480 \frac{\text{kg}}{\text{m}^3} \times \frac{45\ 600}{10^9} \text{ m}^3$$

$$= 0·39 \text{ kg}.$$

1.4. STANDARD TEMPERATURE AND PRESSURE

In general, changes in temperature and pressure affect the volume (hence density) of a substance. It is necessary, therefore, to measure the density of a substance under standard temperature and pressure (s.t.p.) conditions. Standard temperature is taken as 0 °C (273 K) and standard pressure of the atmosphere as 101 325 Pa, say 0·1 MPa, except where the most accurate work is required.

1.5. FORCE AND PRESSURE

1.5.1. Force. Force may be defined as *that which changes,* or tends to change, the state of rest of a body or its uniform motion in a straight line.

Force is not a visible phenomenon, but its effect can be seen and felt. (These effects are considered in later chapters.)

The unit of force is the newton (N) and is that force which, when acting on a mass of one kilogram (1 kg) gives it an acceleration of one metre per second per second (1 m/s²) (see Chapter 9). Note that, expressed in terms of base units, 1 N equals 1 m kg/s².

1.5.2. Pressure. When a compressive force is distributed over a surface area a pressure is set up. Pressure is a measure of concentration of force, and may be expressed

TABLE 1.2   The following table shows typical figures for the densities of some common substances. These values may be affected by factors such as temperature change, purity, etc. Imperial units are included in this table for the purpose of comparison.

Substance	Density ρ			Relative density d
	(kg/m³)	(g/cm³)	(lb/ft³)	
Aluminium	2720	2·72	170	2·72
Brass	8480	8·48	530	8·48
Cadmium	8570	8·57	535	8·57
Chromium	7030	7·03	440	7·03
Coal	1440	1·44	90	1·44
Copper	8790	8·79	550	8·79
Iron (cast)	7200	7·20	450	7·20
Iron (wrought)	7750	7·75	485	7·75
Lead	11 350	11·35	710	11·35
Nickel	8730	8·73	545	8·73
Nylon	1120	1·12	70	1·12
P.V.C.	1360	1·36	85	1·36
Rubber	960	0·96	60	0·96
Steel	7820	7·82	490	7·82
Tin	7280	7·28	455	7·28
Zinc	7120	7·12	445	7·12
Alcohol	800	0·80	50	0·80
Mercury	13 590	13·59	845	13·59
Paraffin	800	0·80	50	0·80
Petrol	720	0·72	45	0·72
Water (pure)	1000	1·00	62·5	1·00
Water (sea)	1020	1·02	64	1·02
Acetylene	1·17	0·001 17	0·073	0·001 17
Air	1·30	0·001 3	0·081	0·001 3
Carbon dioxide	1·98	0·001 98	0·124	0·001 98
Carbon monoxide	1·26	0·001 26	0·079	0·001 26
Hydrogen	0·09	0·000 09	0·006	0·000 09
Nitrogen	1·25	0·001 25	0·078	0·001 25
Oxygen	1·43	0·001 43	0·089	0·001 43

Note: The figures for gases are at 0 °C (273 K), 101 325 Pa (i.e. standard temperature and pressure).

as *normal force per unit area*, where normal means at right-angles to the surface,

$$\text{pressure } p = \frac{\textbf{normal force } F}{\textbf{area } A}.$$

The SI unit of pressure is given the special name of pascal (Pa). In terms of other SI units, $1 \text{ Pa} = 1 \text{ N/m}^2$. Expressed in terms of base units, $1 \text{ Pa} = 1 \text{ kg/(m s}^2)$. It should be noted that

$$1 \text{ MPa} = 1 \text{ MN/m}^2 = 1 \text{ N/mm}^2.$$

EXAMPLE 1.3. The piston of a diesel engine has a diameter of 520 mm. What force is exerted on the piston when the mean pressure of the gas in the cylinder is 0·72 MPa?

SOLUTION

$$\text{Pressure} = \frac{\text{normal force}}{\text{area}},$$

$$\therefore \text{ force on piston} = \text{pressure} \times \text{area}.$$

$$\text{Piston area} = \pi r^2$$
$$= \pi(0\cdot26)^2 = 0\cdot067\ 6\ \pi \text{ m}^2.$$

Hence, force on piston $= 0\cdot72\ \dfrac{\text{MN}}{\text{m}^2} \times 0\cdot067\ 6\ \pi \text{ m}^2$

(where $0\cdot72 \text{ MN/m}^2 = 0\cdot72 \text{ MPa}$)

$$= 0\cdot153 \text{ MN}.$$

EXAMPLE 1.4. A cylindrical oil drum, 0·4 m diameter, 0·6 m high standing on its circular end is filled with oil of 0·8 relative density. If the density of water is 1000 kg/m³ calculate:

(a) the density of the oil,
(b) the total mass of the oil, and
(c) the pressure exerted on the bottom of the drum.

SOLUTION

(a) Density of oil

$$= \text{density of water} \times \text{relative density of oil}$$

$$= 1000\ \frac{\text{kg}}{\text{m}^3} \times 0\cdot8$$

$$= 800 \text{ kg/m}^3\ (= 0\cdot8 \text{ g/cm}^3).$$

TABLE 1.3

Conversion of units (non-SI units are included in this table for the purpose of comparison with SI units)

	SI (international system)	c.g.s. system	Engineers' (or 'gravitational') system	f.p.s. (or 'absolute') system
Length	1 m 0·01 m 0·304 8 m 0·025 4 m	100 cm 1 cm 30·48 cm 2·54 cm	3·281 ft = 39·37 in 0·032 81 ft = 0·3937 in 1 ft = 12 in 0·083 3 ft = 1·0 in	
Mass	1 kg 0·001 kg 14·61 kg 0·453 6 kg	1000 g 1 g 14 610 g 453·6 g	0·068 44 slug 0·000 068 44 slug 1 slug 0·031 06 slug	2·205 lb 0·002 205 lb 32·2 lb 1 lb
Time	1 s	1 s	1 s	1 s
Force	1 N 0·000 01 N 4·45 N 0·138 2 N	100 000 dyn 1 dyn 445 000 dyn 13 820 dyn	0·225 lbf 0·000 002 25 lbf 1 lbf 0·031 06 lbf	7·236 pdl 0·000 072 36 pdl 32·2 pdl 1 pdl
Area	1 m² 0·000 1 m² 0·092 9 m² 0·000 645 m²	10 000 cm² 1 cm² 929 cm² 6·452 cm²	10·76 ft² 0·001 076 ft² = 0·155 in² 1 ft² = 144 in² 0·006 944 ft² = 1 in²	
Volume	1 m³ 0·000 001 m³ 0·028 32 m³ 0·000 016 39 m³	1 000 000 cm³ 1 cm³ 28 320 cm³ 16·39 cm³	35·31 ft³ 0·000 035 31 ft³ 1 ft³ = 1728 in³ 0·000 579 ft³ = 1 in³	
Capacity	1 l (= 1000 cm³) 4·546 l		0·22 gal 1 gal	

(Note that the density and relative density of a substance are numerically equal when the density is expressed in g/cm³.)

(b) Mass of oil = density of oil × volume of drum

$$= 800 \frac{kg}{m^3} \times (\pi \times 0 \cdot 2^2 \times 0 \cdot 6) \, m^3$$

$$= 19 \cdot 2 \, \pi \, kg$$

$$= 60 \cdot 33 \, kg.$$

(c) Pressure $= \dfrac{force}{area}$ (where force $F = mg$)

$$= \frac{60 \cdot 33 \times 9 \cdot 81 \, N}{\pi \times 0 \cdot 2^2 \, m^2} \quad (1 \, N/m^2 = 1 \, Pa).$$

$$= 4708 \cdot 8 \, Pa, \text{ say } 4 \cdot 71 \, kPa.$$

SUMMARY

Density of a substance $(\rho) = \dfrac{\text{mass } (m)}{\text{volume } (V)}$.

Some units of density are: g/cm³, kg/m³.

Relative density (d)

$$= \frac{\text{density of substance}}{\text{density of pure water at } 4 \,°C \, (277 \, K)}$$

$$= \frac{\text{mass of substance}}{\text{mass of equal volume of water}}.$$

Relative density has no units: it is a ratio and is therefore, expressed as a number.

$$\text{Pressure } (p) = \frac{\text{normal force } (F)}{\text{area } (A)}.$$

Some units of pressure are: Pa, MPa.

Force is that which changes, or tends to change, the state of rest or uniform motion of a body.

Some units of force are: N, kN, MN.

EXERCISES 1

1. The density of water is 1000 kg/m³ and nitric acid has a relative density of 1·2.
   The density of nitric acid is...............................kg/m³.

Most relevant sections: 1.3.1 and 1.3.2

U.E.I.

2. (a) The magnitude of a surface is a two-dimensional quantity called..
   (b) The three-dimensional quantity called.................. describes the space occupied by a substance.

Most relevant sections: 1.2.5 and 1.2.6

U.E.I.

3. (a) Force is defined as..
   (b) Pressure can be measured in units of..................

Most relevant sections: 1.5.1 and 1.5.2

U.E.I.

4. Complete the following table:

Material	Density	Relative density
Water	1000 kg/m³	....
Aluminium	2·7 g/cm³	....
Iron	8 t/m³	....

Most relevant sections: 1.3.1 and 1.3.2

U.E.I.

5. Using British Standard (B.S.) symbols and abbreviations, write down correct units of area....................
   force..............................., density...........................

Most relevant sections: 1.2.5, 1.5.1, and 1.3.1

6. Give abbreviations for:
   (a) meganewton ....
   (b) kilonewton per square metre.................................

Most relevant sections: 1.5.1 and 1.5.2

7. A piece of material has a mass m (kg) and a volume V (m³). What will be the density of the material and what are the units of the density?

Most relevant section: 1.3.1

U.L.C.I.

8. A cylindrical tank, 1 m diameter, 2 m high is supported on its circular end and is filled with a liquid of relative density 1·3. If the density of water is 1000 kg/m³ calculate:
   (a) the density of the liquid in kg/m³,
   (b) the total mass of the liquid in kg, and
   (c) the pressure exerted on the end of the tank in kPa.

Most relevant sections: 1.3.1, 1.3.2, and 1.5.2

U.E.I.

9. Distinguish between mass and weight.

Most relevant sections: 1.2.2 and 1.2.4

U.L.C.I.

10. What are units of length, force, and time in the SI?

Most relevant section: Table 1.3

U.L.C.I.

11. A piece of metal has mass m (g) and volume V (cm³). Find expressions for its density and its relative density stating units where appropriate. The mass of 1 m³ of water is 1000 kg.

Most relevant sections: 1.3.1 and **1.3.2**

C.G.L.I.

12. Calculate the mass of 24 cm³ of steel, given that its relative density is 7·8 and that the mass of 1 m³ of water is 1000 kg.

Most relevant sections: 1.3.1 and **1.3.2**

C.G.L.I.

13. Define density and relative density. Determine the relative density of a substance which has a density of 7·12 g/cm³. One cubic metre of water has a mass of 1000 kg.

Most relevant sections: 1.3.1 and **1.3.2**

Y.C.F.E.

14. A rectangular block of cast iron has a volume of 500 cm³ and rests on a horizontal surface. The area of the block in contact with the surface is 1250 mm². If the density of the cast iron is 7·2 g/cm³, calculate the pressure between the block and the surface in MPa.

Most relevant sections: 1.3.1 and **1.5.2**

E.M.E.U.

ANSWERS TO EXERCISES 1

1. 1200 kg/m³. **4.** 1, 2·7, 8. **7.** $\rho = m/V$ (kg/m³). **8.** (a) 1300 kg/m³, (b) 2043 kg, (c) 255 kPa. **11.** $\rho = m/V$ (g/cm³): $d = m/V$ (no units). **12.** 0·187 kg. **13.** 7·12. **14.** 28·2 MPa.

2 Forces

2.1. FORCE

Force has been defined as that which changes, or tends to change, the state of rest of a body or its uniform motion in a straight line.

There are various kinds of forces, including the forms commonly known as *direct forces*, *hydraulic forces*, and *magnetic forces*.

In the present chapter, the effects of direct forces will be considered. Two examples of direct forces are the pull in the towrope between a tug and a ship being towed, and the push exerted by a bulldozer when shifting earth.

2.1.1. Characteristics of force. To describe a force completely, it is necessary to know the following details:

1. *Magnitude*—which means the size of the force, measured in force units (e.g. 3 N, 4 kN, 5 MN).
2. *Direction*—which means the line of action of the force (e.g. horizontal, vertical, or otherwise).
3. *Sense*—meaning which way along the line of action the force is acting (e.g. left or right, up or down).

2.1.2. Graphical representation of force. The above-mentioned characteristics define a quantity called a *vector*. (A vector can be shown on a drawing by a straight line of definite length and direction, which represents to scale the magnitude and direction of a physical quantity such as a force. An arrowhead is used to indicate sense.)

Consider a force of 10 kN pulling away from a point O on a body. The force is inclined at 20° to the horizontal, and acting towards the right.

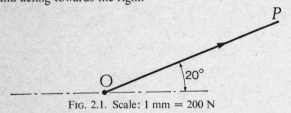

FIG. 2.1. Scale: 1 mm = 200 N

The line OP in Fig. 2.1 is drawn to a scale of 1 mm = 200 N (any convenient scale will do) so that the length of the line represents the magnitude of the force. The direction of the force is represented by the line being drawn at 20° to the horizontal, while the arrowhead shows the sense as pulling away from the point of application O.

2.1.3. Resultant. When two or more forces act on a body, each force has some effect on the body. The total effect produced is a combination of the separate effects of the forces acting on the body and could also be produced by some single force.

This single force is known as the *resultant*, and is that force which will replace the existing system of forces acting on a body and produce the same effect upon the body. On a diagram the resultant has a double arrow on its vector.

2.1.4. Vector addition. The magnitude, direction, and sense of the resultant are determined graphically by the addition of the vectors.

Fig. 2.2 shows two forces F_1 and F_2 acting at a point O. By arranging the vectors end-to-end as shown, the resultant R is found by connecting the start of the first vector to the finish of the last vector drawn. In Figs 2.2 (a) and (b), the vectors should be superimposed but are shown side by side for clarity.

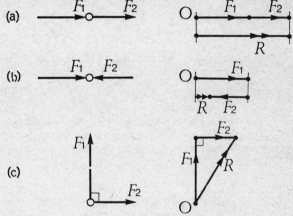

FIG. 2.2. Vector addition

2.1.5. Equilibrium. If a force, or system of forces, is acting on a body, and is balanced by some other force, or system of forces, then the body is said to be in *equilibrium*. For example, a stationary body is in equilibrium.

2.1.6. Equilibrant. The *equilibrant* of a system of forces is that force which, when added to the system, produces equilibrium. This is shown in Fig. 2.3.

It has been seen that the resultant is the single force which will replace an existing system of forces and produce the same effect. It therefore follows that if the equilibrant is to produce equilibrium is must be equal in magnitude and direction, but opposite in sense to the resultant.

2.2. PARALLELOGRAM OF FORCES

If two forces acting at a given point are represented in magnitude and direction by adjacent sides of a parallelo-

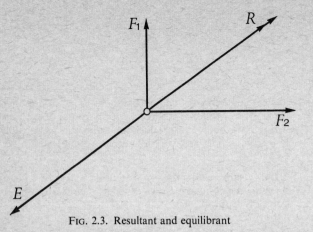

FIG. 2.3. Resultant and equilibrant

same magnitude and direction as *R* but is of opposite sense.

EXAMPLE 2.1. A right-handed turning tool cuts a bar of metal as shown in Fig. 2.5. The force required to maintain the cut is 100 N and the feed force is 350 N. Find, graphically, the resultant of these two forces.

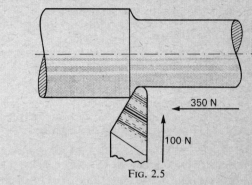

FIG. 2.5

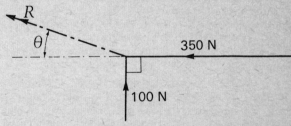

Space diagram   Force (vector) diagram

FIG. 2.4. Parallelogram of forces

FIG. 2.6. Space diagram

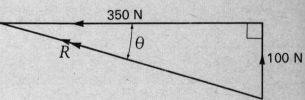

FIG. 2.7. Force (vector) diagram. Scale: 1 mm = 5 N

gram, the resultant of these two forces will be represented by the diagonal of the parallelogram at that point.

Hence, in Fig. 2.4 the resultant of the two forces F_1 and F_2 is given by the diagonal *R*. It should be noted that the same construction gives the equilibrant *E* which has the

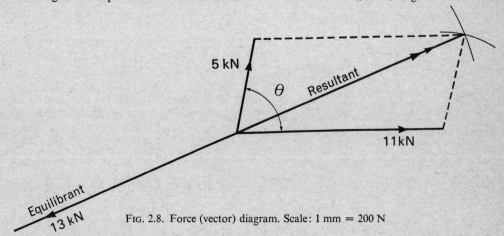

FIG. 2.8. Force (vector) diagram. Scale: 1 mm = 200 N

SOLUTION

The two forces act at right-angles to one another as shown in the space diagram (Fig. 2.6).

In this example the resultant is determined by the method of vector addition (Fig. 2.7).

Magnitude of resultant force $R = 364$ N.

Direction $\theta = 16°$ above the horizontal and acting towards the left.

EXAMPLE 2.2. (a) Find, graphically, at what angle two forces of 5 kN and 11 kN must be inclined if they are balanced by a force of 13 kN.

(b) The resultant of forces 140 N and 75 N is at right angles to the smaller force. Draw a diagram to show the resultant and state its magnitude.

U.E.I.

SOLUTION

This problem has been solved by applying the parallelogram of forces. See Figs 2.8 and 2.9.

(a) The angle θ between the two forces must be 79°.

(b) Magnitude of resultant force $R = 115$ N.

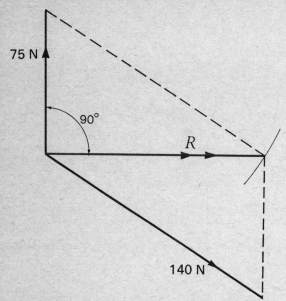

FIG. 2.9. Force (vector) diagram. Scale: 1 mm = 2 N

2.3. THREE FORCES IN EQUILIBRIUM

If three coplanar forces (i.e. three forces acting in the same plane) are in equilibrium:

(1) they may be represented in magnitude and direction by the sides of a triangle drawn to scale, and

(2) they are concurrent (i.e. the lines of action of the forces pass through a common point).

The above statement is known as the *triangle of forces law*.

Fig. 2.10 shows three coplanar forces acting on a body which is in equilibrium. The line of action of each force

is produced to meet at a common point O, called the point of concurrency.

The corresponding force diagram is shown by Fig. 2.11, which must be drawn accurately to a suitable scale so that the sides of the triangle represent the magnitude and direction of the three forces. The sense of each force is indicated by the arrowhead placed on the respective vector.

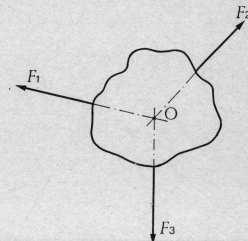

FIG. 2.10. Space diagram. Three forces in equilibrium

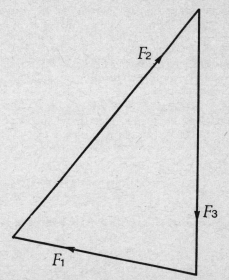

FIG. 2.11. Triangle of forces. Force (vector) diagram

2.4. BOW'S NOTATION

Bow's notation is a convenient system of lettering the forces for reference purposes when there are three or more forces to be considered.

Capital letters are placed in the space between forces

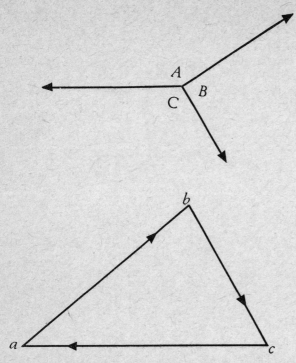

FIG. 2.12. Bow's notation

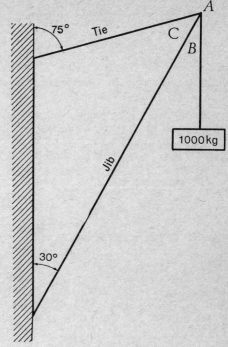

FIG. 2.13. Space diagram

in a clockwise direction as shown in Fig. 2.12. Any force is then referred to by the letters in its adjacent spaces. The vectors representing the forces are given the corresponding small letters. Thus the forces, *AB*, *BC*, and *CA* are represented by the vectors *ab*, *bc*, and *ca* respectively. This will apply to any number of forces and their corresponding vectors.

EXAMPLE 2.3. Fig. 2.13 shows a simple wall crane. Determine, graphically, the forces in the jib and tie when a mass of 1000 kg is suspended as shown. (Take g as 10 m/s².)

SOLUTION

The three forces are in equilibrium so that the triangle of forces law may be applied as in Fig. 2.14. In this example Bow's notation is used.

The gravitational force on a mass of 1000 kg is given by

$$F = mg.$$

(See sections 9.1–9.3 for an explanation of this formula.) In this case, force $F = 1000 \text{ kg} \times 10 \text{ m/s}^2$

$$= 10\ 000 \text{ N}.$$
$$= 10 \text{ kN}.$$

Force in jib = 13·7 kN.

Force in tie = 7·1 kN.

Note that in order to keep the arrangement in equilibrium, the force in the jib is 'pushing' upwards and the force in the tie is 'pulling' downwards. This means that

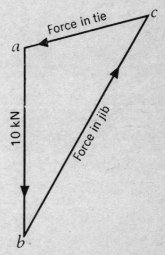

FIG. 2.14. Force (vector) diagram. Scale: 1 mm = 0·2 kN

the jib must be a rigid member, while the tie may be either rigid of flexible such as a chain or wire rope.

EXAMPLE 2.4. A steel ladder weighing 400 N hangs from a hinge at its top end and is pulled aside from the vertical by a horizontal force of 200 N acting at the bottom end of the ladder. Find graphically the direction of the reaction at the hinge assuming the weight of the ladder acts through its mid-point.　　　　　U.E.I.

SOLUTION

Note: The three forces are in equilibrium and are concurrent. The weight of the ladder acts vertically downwards. The direction of the reaction at the hinge will be through the point of concurrency and the centre of the hinge. (see Figs 2.15 and 2.16).

Direction of the reaction at the hinge, $\theta = 63\frac{1}{2}°$.

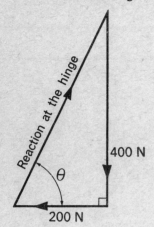

FIG. 2.15. Force (vector) diagram. Scale: 1 mm = 8 N

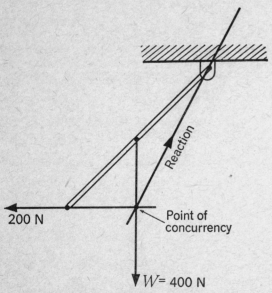

FIG. 2.16. Space diagram

2.5. POLYGON OF FORCES

Often, in engineering problems, it is required to find the resultant or equilibrant of three or more forces, in which case an extension of the parallelogram law is used and is known as the polygon of forces rule.

The *polygon of forces* rule states that *if three or more coplanar forces acting on a body are in equilibrium they can be represented by the sides of a polygon taken in order*.

EXAMPLE 2.5. Three coplanar forces *A*, *B*, and *C* act through the same point. Force *A* = 8 kN. Force *B* = 11 kN and acts at 60° to *A*. Force *C* = 6 kN and acts at 225° to *A*, both angles being measured anticlockwise from *A*.

Determine the magnitude and direction of the equilibrant of *A*, *B*, and *C*.

U.E.I.

SOLUTION

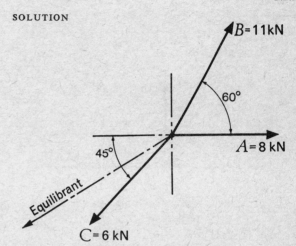

FIG. 2.17. Space diagram

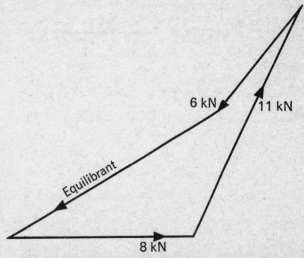

FIG. 2.18. Force (vector) diagram. Scale: 1 mm = 160 N

The space diagram (Fig. 2.17) is drawn with correct angles between the forces.

The force diagram (Fig. 2.18) is drawn by the method of vector addition and the polygon completed by adding the equilibrant.

All arrowheads point nose to tail round the polygon.

The equilibrant has a magnitude of 10·5 kN and acts in a direction 210° anticlockwise to the line of action of the 8 kN.

EXAMPLE 2.6. A bolt is acted upon by four coplanar forces which have the following magnitudes and directions: (250 N, 0°), (400 N, 90°), (500 N, 135°), and (350 N, 240°).

Find the magnitude and direction of the resultant of the forces acting on the bolt.

SOLUTION

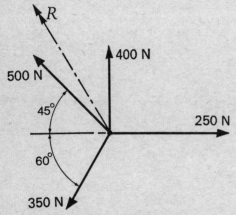

FIG. 2.19. Space diagram

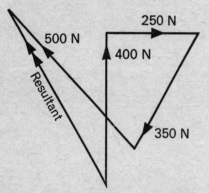

FIG. 2.20. Force (vector) diagram. Scale: 1 mm = 10 N

Draw the space and force diagrams, Figs 2.19 and 2.20, as in the previous example.

Note: All angles are measured from the horizontal in an anticlockwise direction.

The resultant has a magnitude of 500 N and acts in a direction 120° anticlockwise to the line of action of the 250 N *or* (500 N, 120°).

In these problems Bow's notation, explained earlier in the chapter, may be used to advantage in referring to the forces.

To introduce Bow's notation to the above problem the following procedure is adopted.

Insert the resultant in its approximate position in the space diagram (Fig. 2.21) and letter the spaces between the forces with capital letters as shown.

In the force diagram (Fig. 2.22) the forces can now be referred to as follows:

400 N *ab*, 250 N *bc*, 350 N *cd*, 500 N *de*,

Resultant *ae*.

Note: If the equilibrant was required it would be referred to as *ea* equal and opposite to the resultant *ae*.

Arrowheads need not be used when this notation is adopted.

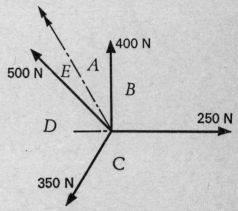

FIG. 2.21. Space diagram

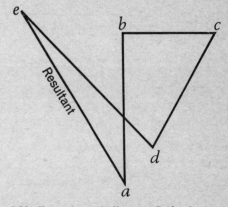

FIG. 2.22. Force (vector) diagram. Scale: 1 mm = 10 N

2.6. RESOLUTION OF FORCES

Graphical solutions to problems involving forces as illustrated in the preceding section are accurate enough for many engineering problems.

However, sometimes it is necessary to provide a more accurate result in which case a mathematical method of solution is essential.

This method of solution is known as *resolution of forces*.

Consider a force F acting on a rivet A as shown in Fig. 2.23. The force F may be replaced by two forces P and Q, acting at right-angles to each other, which together have the same effect on the rivet.

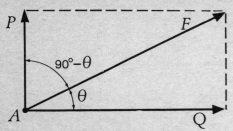

FIG. 2.23. Resolution of forces

From a knowledge of elementary trigonometry,

$$\frac{Q}{F} = \cos \theta,$$

$$\therefore Q = F \cos \theta. \tag{i}$$

$$\frac{P}{F} = \cos (90° - \theta),$$

$$\therefore P = F \cos (90° - \theta), \text{ but } \cos (90° - \theta) = \sin \theta$$

$$\therefore P = F \sin \theta. \tag{ii}$$

From the diagram:

$$P = F \sin \theta.$$
$$Q = F \cos \theta.$$

Hence the one force F has been resolved or split into two equivalent forces of magnitudes $F \cos \theta$ and $F \sin \theta$ at right-angles to each other.

$F \cos \theta$ is known as the *horizontal component* of F.
$F \sin \theta$ is known as the *vertical component* of F.

EXAMPLE 2.7. A force of 200 N is applied along the length of a chisel held at 60° to the horizontal. What force is exerted at the end of the chisel (a) horizontally (b) vertically.

U.E.I.

SOLUTION

See Fig. 2.24.

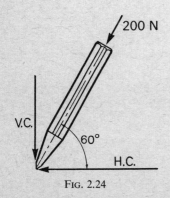

FIG. 2.24

(a) Horizontal component $= 200 \cos 60° = 200 \times 0.5$ N
$$= 100 \text{ N}.$$
(b) Vertical component $= 200 \cos 30°$ or
$$200 \sin 60° \text{ N}$$
$$= 200 \times 0.866 = 173.2 \text{ N}.$$

From this example it can be seen that the horizontal component of the 200 N influences the length of the cut and the vertical component determines the depth of the cut.

2.6.1. Determination of resultant or equilibrant (resolution method). This method of solution will be easier to follow by means of worked examples.

Consider the two examples previously solved by a graphical method.

EXAMPLE 2.8. Three coplanar forces A, B, and C act through the same point. Force $A = 8$ kN, force $B = 11$ kN and it acts at 60° to A. Force $C = 6$ kN and acts at 225° to A, both angles being measured anticlockwise from A.

Determine the magnitude and direction of the equilibrant of A, B, and C.

U.E.I.

SOLUTION

Each force will be resolved into two forces, one along the vertical axis and one along the horizontal axis.

The normal algebraic sign convention will be used: i.e. above the origin V is positive and below negative. H is positive to the right of the origin and negative to the left. By this convention sines and cosines of acute angles only will be required. See Fig. 2.25.

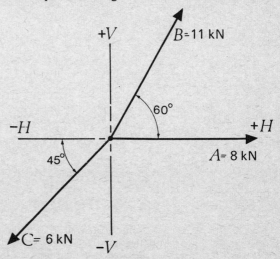

FIG. 2.25

Magnitude of force	Horizontal component	Vertical component
8 kN	+8 kN (→)	0
11 kN	+11 cos 60° kN (→)	+11 sin 60° kN (↑)
6 kN	−6 cos 45° kN (←)	−6 sin 45° kN (↓)

$$\begin{aligned} \text{Total horizontal component} &= 8 + 11 \cos 60° \\ &\quad - 6 \cos 45° \text{ kN} \\ &= 8 + 5 \cdot 5 - 4 \cdot 24 \text{ kN} \\ &= 9 \cdot 26 \text{ kN} (\rightarrow). \end{aligned}$$

Since 9·26 is a positive number, the force will act from left to right.

$$\begin{aligned} \text{Total vertical component} &= 0 + 11 \sin 60° - 6 \sin 45° \text{ kN} \\ &= 0 + 9 \cdot 53 - 4 \cdot 24 \text{ kN} \\ &= 5 \cdot 29 \text{ kN} (\uparrow). \end{aligned}$$

Since 5·29 is a positive number, the force will act upwards.

The three original forces have now been reduced to two forces acting at right-angles to each other.

The magnitude of the resultant R, or equilibrant, may be obtained by using Pythagoras' theorem on the right-angled triangle obtained. This is shown in Fig. 2.26.

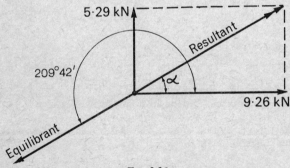

Fig. 2.26

$$\begin{aligned} R^2 &= 5 \cdot 29^2 + 9 \cdot 26^2 \\ &= 27 \cdot 99 + 85 \cdot 73 = 113 \cdot 7, \\ R &= 10 \cdot 6. \end{aligned}$$

Magnitude of equilibrant = 10·6 kN.

From the right-angled triangle, the angle α the equilibrant makes with the horizontal direction may be calculated by use of simple trigonometry.

$$\tan \alpha = \frac{5 \cdot 29}{9 \cdot 26} = 0 \cdot 571,$$

$$\alpha = 29° \, 42'.$$

∴ equilibrant makes an angle of (180° + 29° 42′) with the line of action of the 8 kN.

The equilibrant has a magnitude of 10·61 kN and acts at an angle of 209° 42′ with the 8 kN.

Compare the answers with those obtained graphically in Example 2.5.

EXAMPLE 2.9. A bolt is acted upon by four coplanar forces which have the following magnitudes and directions: (25 N, 0°), (40 N, 90°), (50 N, 135°), and (35 N, 240°).

Find the magnitude and direction of the resultant of the forces acting on the bolt (see Fig. 2.27).

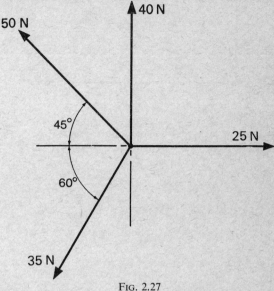

Fig. 2.27

SOLUTION

Note: In this example a knowledge of cosine ratios of angles greater than 90° is assumed.

Each force will be resolved into two forces, one along the vertical axis in the direction of the 40 N and one along the horizontal axis in the direction of the 25 N.

Magnitude of force (N)	Horizontal component (N)	Vertical component (N)
25	25 (→)	0
40	0	40 (↑)
50	50 cos 135° (→)	50 cos 45° (↑)
35	35 cos 120° (→)	35 cos 150° (↑)

Total horizontal component

$$\begin{aligned} &= 25 + 0 + 50 \cos 135° + 35 \cos 120° \text{ N} \\ &= 25 + 0 + (50 \times -0 \cdot 707 \, 1) + (35 \times -0 \cdot 5) \text{ N} \\ &= 25 - 35 \cdot 36 - 17 \cdot 5 \text{ N} \\ &= -27 \cdot 86 \text{ N}. \end{aligned}$$

The negative sign indicates that the horizontal component acts from right to left.

Horizontal component = 27·86 N (←).

Total vertical component

$$\begin{aligned} &= 0 + 40 + 50 \cos 45° + 35 \cos 150° \text{ N} \\ &= 0 + 40 + (50 \times 0 \cdot 707 \, 1) + (35 \times -0 \cdot 866) \text{ N} \\ &= 0 + 40 + 35 \cdot 35 - 30 \cdot 31 \text{ N} \\ &= 45 \cdot 04 \text{ N}. \end{aligned}$$

Vertical component = 45·04 N (↑)

The four forces have now been reduced to two forces as shown in Fig. 2.28.

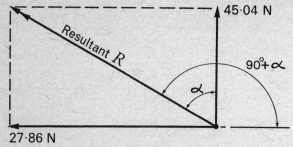

27·86 N

FIG. 2.28

Using Pythagoras' theorem,

$$R^2 = (27·86)^2 + (45·04)^2,$$
$$R^2 = 776·1 + 2029,$$
$$R^2 = 2805,$$
$$R = 52·96.$$

Magnitude of resultant $= 52·96$ N.

To determine direction α,

$$\tan \alpha = \frac{27·86}{45·04} = 0·618\,5,$$
$$\alpha = 31°\,44'.$$

Resultant has a magnitude of 52·96 N and acts at an angle of 31° 44′ with the direction of the 40 N, or (52·96 N, 121° 44′).

2.7. EQUILIBRIUM ON A SMOOTH INCLINED PLANE

The term 'smooth' implies that the effects of friction may be ignored.

The body is kept in equilibrium on the plane by the action of three forces as shown in Figs 2.29 and 2.33. These forces are:

(1) The weight W of the body acting vertically downwards.
(2) The reaction R of the plane to the weight of the body. R is termed the *normal reaction*, *normal* in this sense meaning *at right-angles to*.
(3) The force P acting in some suitable direction required to prevent the body sliding down the plane.

The forces P and R are dependent on three quantities:

(a) the angle of inclination of the plane;
(b) the magnitude of W; and
(c) the inclination of the force P to the plane.

It is, therefore, possible to express the magnitudes of both forces P and R in terms of W and functions of θ and α.

Two positions of equilibrium will be considered:
(I) force P acting parallel to the plane;
(II) force P acting at an angle α to the plane.

2.7.1. *Case I:* Force P acting parallel to the plane.

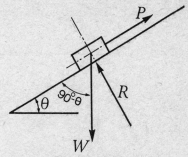

FIG. 2.29. Space diagram

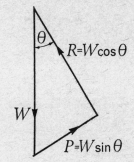

FIG. 2.30. Force (vector) diagram

(a) *Resolution method*

W may be resolved into two forces at right-angles to each other, one along the plane and one at right-angles to the plane.

Referring to Fig. 2.29, force component at right-angles to the plane $= W \cos \theta$, and force component parallel to the plane $= W \sin \theta$.

Equating forces $W \cos \theta = R$, (i)
$W \sin \theta = P$. (ii)

(b) *Graphical method*

The forces R and P may be found by the application of the triangle of forces rule, as shown in Fig. 2.30.

EXAMPLE 2.10. A truck of mass 500 kg is on a smooth plane inclined at 20° to the horizontal. What force parallel to the inclined plane is necessary to prevent the truck from moving? Take g as 10 m/s².

C.G.L.I.

SOLUTION

The gravitational force exerted on a mass of 500 kg is given by $F = mg$.

Then weight, $W = 500 \times 10$ N
$= 5000$ N
$= 5$ kN.

Resolving forces parallel to the plane in Fig. 2.31:

$$W \sin \theta = P,$$
$$5 \sin 20° = 5 \times 0\cdot342 = 1\cdot71,$$
$$P = 1\cdot71 \text{ kN}.$$

Graphical method of solution:
Apply the triangle of forces rule as illustrated in Fig. 2.32.

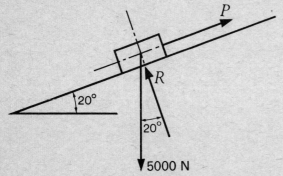

FIG. 2.31. Space diagram

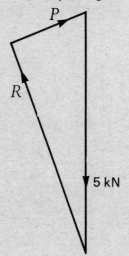

FIG. 2.32. Force (vector) diagram. Scale: 1 mm = 75 N

2.7.2. *Case II:* Force *P* acting at an angle α to the plane.
(a) *Resolution method* shown in Fig. 2.33.
Resolving forces along the plane,
$$P \cos \alpha = W \sin \theta. \qquad (i)$$
Resolving forces at right-angles to the plane,
$$P \sin \alpha + R = W \cos \theta. \qquad (ii)$$

(b) *Graphical method*
The unknown forces may be obtained by the application of the triangle of forces rule (see Fig. 2.34).

EXAMPLE 2.11. A frictionless trolley of mass 10 kg is held by a cord on a smooth inclined plane. The force in the cord acts upwards at 30° to the plane and the plane is inclined at 20° to the horizontal. Take *g* as 10 m/s².

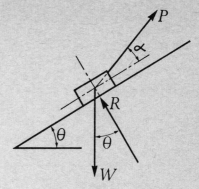

FIG. 2.33. Space diagram

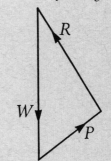

FIG. 2.34. Force (vector) diagram

(a) Determine graphically or by calculation the values of the force in the cord and the reaction to the plane.
(b) Determine the component of the force in the cord acting parallel to the plane.

Y.C.F.E.

Note: The term 'frictionless trolley' means the effects of friction may be ignored. In practice friction is always present.

SOLUTION
Weight $= mg = 100$ N.

Resolution of forces:
(a) Resolving forces along the plane,

$$P \cos 30° = 100 \sin 20° \qquad (i)$$
$$P = \frac{100 \sin 20°}{\cos 30°}$$
$$= \frac{34\cdot2}{0\cdot866}$$
$$= 39\cdot4.$$

Force in the cord $= 39\cdot4$ N.

Resolving forces at right-angles to the plane,

$$P \sin 30° + R = 100 \cos 20°, \qquad (ii)$$
$$39\cdot4 \times 0\cdot5 + R = 100 \times 0\cdot9397,$$
$$R = 93\cdot97 - 19\cdot7$$
$$= 74\cdot27.$$

Reaction to the plane $= 74\cdot27$ N.

(b) Component of the force in the cord acting parallel

to the plane = $P \cos 30°$ N

$= 39\cdot4 \times 0\cdot866$ N

$= 34\cdot2$ N.

Graphical method of solution:

Apply triangle of forces rule. Draw space diagram Fig. 2.35.

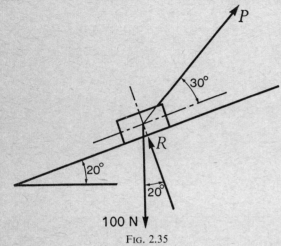

FIG. 2.35

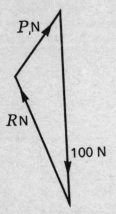

FIG. 2.36. Force (vector) diagram. Scale: 1 mm = 2 N

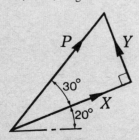

FIG. 2.37. Force (vector) diagram. Scale: 1 mm = 1 N

(a) From Fig. 2.36, $P = 42$,

$R = 75$.

Force in the cord = 42 N.

Reaction to the plane = 75 N.

(b) In Fig. 2.37 Y represents the component of the force in the cord at right-angles to the plane.

X represents the component of the force in the cord parallel to the plane.

$$X = 33\cdot8$$

Component of the force in the cord acting parallel to the plane = $33\cdot8$ N.

SUMMARY

Vector quantities are those which possess both magnitude and direction. They may be completely represented by a straight line (magnitude and direction) and an arrowhead (sense).

Force is a vector quantity.

The resultant of a system of forces is that single force which will replace the existing system of forces acting on a body and produce the same effect upon the body.

The equilibrant of a system of forces is that force which, when added to the system produces equilibrium. It is equal and opposite to the resultant.

Parallelogram of forces:

If two forces acting at a given point are represented in magnitude and direction by the adjacent sides of a parallelogram, the resultant of these two forces will be represented by the diagonal of the parallelogram at that point.

Triangle of forces:

If three coplanar forces acting at a point can be represented in magnitude and direction by the sides of a triangle the forces will be in equilibrium.

Bow's notation is a system of lettering the space and vector diagrams.

Polygon of forces:

If any number of coplanar forces acting at a point can be represented in magnitude and direction by the sides of a polygon the forces will be in equilibrium.

Resolution of forces:

The resolved part of a force F in a direction making an angle θ with it is $F \cos \theta$.

EXERCISES 2

1. (a) A quantity which has magnitude and direction is a...quantity.

(b) Give one example of such a quantity.

Most relevant section: 2.1.2

U.E.I.

2. If two equal forces act at a point, one due north and the other due east, state precisely the direction of their resultant.

Most relevant section: **2.1.3**

U.E.I.

3. What is meant by three concurrent coplanar forces?

Most relevant section: **2.3**

U.L.C.I.

4. Two tugs are towing a ship at constant speed against a total resistance of 50 kN. The angles between the tow ropes and the direction of motion are 30° and 60° respectively. What is the tension in each of the tow ropes?

Most relevant section: **2.2**

U.E.I.

5. (a) Find graphically the magnitude and direction of the resultant of the two forces shown in Fig. 2.38.
(b) Find graphically, the magnitude and direction of the force which must be added to the force of magnitude 40 N shown in Fig. 2.39, so as to produce the resultant of magnitude 60 N at 30° to the 40 N force.

Most relevant section: **2.2**

U.E.I.

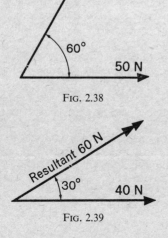

Fig. 2.38

Fig. 2.39

6. In Fig. 2.40, the uniform bar *OA* has a mass of 3 kg. It is pivoted at *O* and supported at 30° to the horizontal by a cord *AB* in the position shown. Find, by drawing a triangle of forces, the tension in the cord *AB* and the magnitude and direction of the reaction at the pivot *O*.
The weight of the bar may be assumed to act through its mid point. Take g as 9·81 m/s².

Most relevant section: **2.3**

U.E.I.

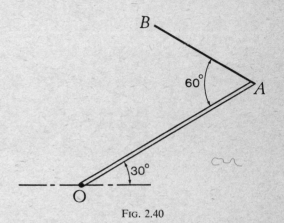

Fig. 2.40

7. A jib crane is arranged as shown in Fig. 2.41. Find the forces in members *AB* and *BC* and state whether they are in tension or compression. Take g as 9·81 m/s².

Most relevant section: **2.3**

U.L.C.I.

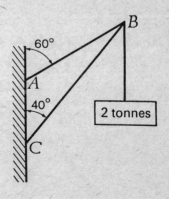

Fig. 2.41

8. Three forces of magnitude 3 kN, 3·5 kN, and 4·25 kN, respectively, act at a point as shown in Fig. 2.42. Determine the angles, measured in a clockwise direction from the 3 kN force, at which the other two forces must act in order that the three forces shall be in equilibrium.

Most relevant section: **2.3**

U.L.C.I.

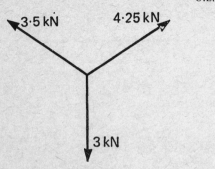

Fig. 2.42

9. A force of magnitude 50 N acts on a body. If the force acts in a vertical plane, making an angle of 30° with the horizontal, find the horizontal and vertical components of the force.

Most relevant section: **2.6**

E.M.E.U.

10. A body of mass 20 kg rests on a smooth surface inclined at an angle of 15° to the horizontal. Determine (a) the force acting parallel to the inclined surface required to prevent the body from slipping; and (b) the normal reaction between the body and the surface.
Take the local gravitational acceleration g as 10 m/s².

Most relevant sections: **2.3** and **2.7.1**

U.E.I.

11. Three coplanar forces pull away from a point in a certain mechanism. The forces have the following magnitudes and directions:

80 N, 0° horizontally to right;
110 N, 60° anticlockwise from the 80 N force;
60 N, 225° anticlockwise from the 80 N force.

Find, graphically or by calculation, the magnitude and direction of the resultant of these three forces. Resolve the resultant into horizontal and vertical components.

Most relevant sections: **2.3** and **2.6**

U.E.I.

12. In Fig. 2.43, a machine M of mass 200 kg is lifted steadily by a cable that passes over a frictionless pulley D and is pulled at E in a direction at 45° to the horizontal. The pulley is suspended from two cables AB and BC, both of which are inclined at 30° to the horizontal. All the cables lie in the same vertical plane. Find (a) the tension in each of the cables AB and BC; (b) the inclination to the vertical of the suspending link BD. Take g as 9·81 m/s².

Most relevant sections: **2.2** and **2.3**

C.G.L.I.

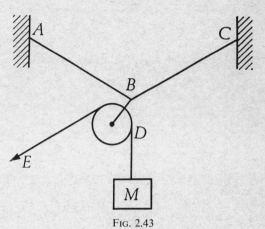

Fig. 2.43

13. A joint in a framework is in equilibrium under the action of four forces which are (200 N, 90°); (80 N, 0°); X and Y. The lines of action of X and Y are respectively −60° and +225°.

Determine graphically or by calculation the magnitude and direction of the forces X and Y.

Calculate the single force in magnitude and direction which would replace X and Y.

Most relevant sections: **2.5** and **2.6**

Y.C.F.E.

14. Describe briefly an experiment which may be carried out to verify the triangle of forces.

An engine has a mass of 90 kg and is lifted by means of two eye bolts 1 m apart, equidistant from the centre of gravity. Two chains, each 1·5 m long, are attached at one end to the eye bolts and at the other to a central lifting hook.

Find the tension in the chains in newtons. Take g as 9·81 m/s².

Most relevant section: **2.3**

E.M.E.U.

15. The wall crane shown in Fig. 2.44 consists of a horizontal jib *AB* supported by a tie bar *BC*.

A mass *m* of 150 kg is suspended by a rope passing over a small freely running pulley at *B*, the free end of the rope being inclined at 30° to the vertical as shown.

Determine the resultant force applied by the rope at *B*, and hence, the magnitude and natures of the forces in the jib and the tie.

Neglect the weights of the jib and tie. Take *g* as 9·81 m/s².

Most relevant sections: **2.2** and **2.3**

U.E.I.

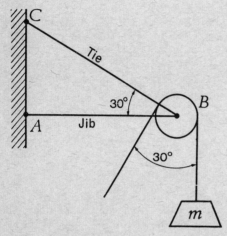

FIG. 2.44

16. A body of mass 20 kg rests on a smooth surface inclined at an angle of 15° to the horizontal. Determine by a graphical method (a) the force acting along the plane, required to prevent the body from slipping, (b) the normal reaction between the body and the plane. Take *g* as 9·81 m/s².

Most relevant sections: **2.7.1** and **2.3**

U.L.C.I.

17. Four forces act on a body as shown in Fig. 2.45. Determine by a graphical method the magnitude of the resultant force and the angle it makes with force *A*.

Most relevant section: **2.5**

U.L.C.I.

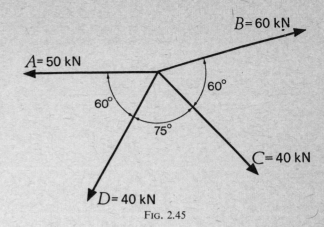

FIG. 2.45

18. A simple symmetrical chain sling is used to lift a forging as shown in Fig. 2.46 below.

(a) If the tension in the portions of the sling marked *B* and *C* are limited to 8000 N, and the angle θ is 60° calculate the maximum force that can be exerted by lifting at *A*.

(b) Taking *g* as 9·81 m/s², determine the maximum mass of the forging in tonnes (t). (The effect of gravity on a mass of *m* (kg) is *mg* (N) and 1 t = 1000 kg).

Most relevant section: **2.3**

C.G.L.I.

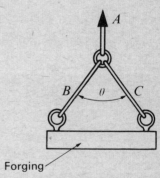

Forging

FIG. 2.46

19. Fig. 2.47 shows a trapdoor being held open at 45° to the vertical by a force *F*. The trapdoor exerts a force of 400 N and this force can be assumed to act at the centre. Determine

(a) the magnitude of the force *F*; and
(b) the magnitude and direction of the hinge reaction *R*.

Most relevant section: **2.3**

C.G.L.I.

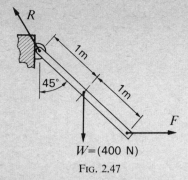

Fig. 2.47

20. Fig. 2.48 shows a forging being lifted by a symmetrical sling.

(a) If the tension in each of the parts *A* and *B* of the sling is limited to 10 kN, find the maximum lifting force *F*.

(b) If the angle of the sling was changed from 60° to 120° for the same sling, what would then be the maximum value of *F*?

Most relevant section: **2.3**

U.E.I.

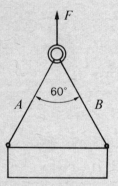

Fig. 2.48

21. Fig. 2.49(a) shows diagrammatically the elements of a lifting crane. The lifting rope, after moving over the pulley, travels horizontally. Fig. 2.49(b) shows the loading at the pulley axle.

Determine by drawing or by calculation:

(a) the magnitude and direction of the load *L*;
(b) the forces in members *X* and *Y* denoted by F_x and F_y. Take *g* as 10 m/s².

Most relevant sections: **2.2** and **2.3**

C.G.L.I.

22. Fig. 2.50 shows the connecting rod, piston, and cylinder walls of an engine. The piston has an effec-

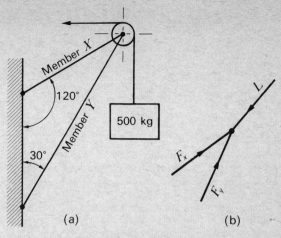

Fig. 2.49

tive area of 8000 mm² and the pressure acting on the piston is 400 kPa.

(a) Calculate the force *P* (kN) acting on the piston.
(b) Determine, by means of a force diagram, the thrust *F* in the connecting rod and the reaction of the cylinder wall *R* when the connecting rod is at the angle shown.

Most relevant sections: **1.5** and **2.3**

U.E.I.

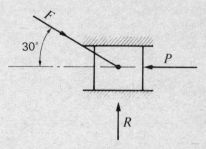

Fig. 2.50

ANSWERS TO EXERCISES 2

2. North-east. **4.** 40 kN, 30 kN. **5.** (a) 70 N, 22° anticlockwise from 50 N, (b) 33 N, $68\frac{1}{2}°$ anticlockwise from 40 N. **6.** 14·7 N, 25·5 N, 60° anticlockwise from horizontal. **7.** $AB = 37\cdot3$ kN (tension), $BC = 50$ kN (compression). **8.** $98\frac{1}{2}°$; $234\frac{1}{2}°$. **9.** 43·3 N; 25 N. **10.** (a) 51·76 N, (b) 193·2 N. **11.** 106 N, 29° 42′, 92·6 N, 52·9 N. **12.** (a) $AB = 2\cdot57$ kN, $BC = 4\cdot17$ kN, (b) $22\frac{1}{2}°$. **13.** 87 N, 300°, 175 N, 225°, 215·4 N, 68° 12′. **14.** 468·2 N. **15.** 2·85 kN, 5·49 kN (compressive), 5·49 kN (tensile). **16.** (a) 51 N, (b) 189·3 N. **17.** 50 kN, $108\frac{1}{2}°$ anticlockwise from *A*. **18.** (a) 13·86 kN, (b) 1·415 t. **19.** (a) 200 N, (b) 450 N, 27° to vertical. **20.** (a) 17·32 kN, (b) 10 kN. **21.** (a) 7·08 kN, 45°, (b) 3·55 kN, 3·9 kN. **22.** (a) 3·2 kN, (b) 3·7 kN, 1·85 kN.

3 Force and turning effect

3.1. TURNING EFFECT

In Figs 3.1 and 3.2 two spanners of length 150 mm and 300 mm are acted upon by hand forces of magnitude 30 N and 15 N respectively.

In each case the line of action of the force is at right angles to the length of the spanner and the point of application is such as to produce a clockwise rotation of the nut.

Experience shows that in practice the 300 mm spanner would require less effort than the 150 mm spanner to produce the same turning effect. It can be seen, that by doubling the length of the spanner the hand force required to produce the same turning effect is halved.

Now consider Fig. 3.3, in which the line of action of the hand force is inclined at an angle of 60° with the length of the spanner.

In this case only part of the applied force is effective in producing a rotation of the spanner. Hence the turning effect of the force is reduced. For purposes of calculation, it is more convenient to consider that the reduction in turning effect is due to a reduction in the effective length of the spanner. This reduced length of spanner is equal to the perpendicular distance from the centre of the nut to the line of action of the 30 N force.

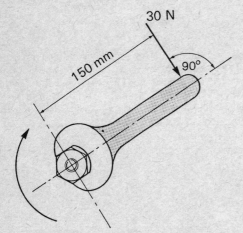

FIG. 3.1. Turning effect of force

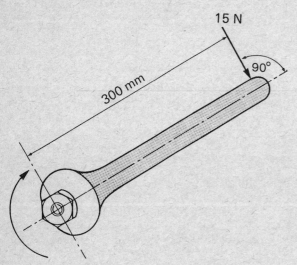

FIG. 3.2. Turning effect of force

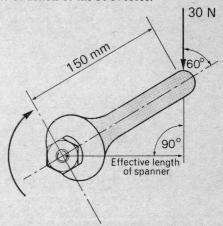

FIG. 3.3. Turning effect of force

3.2. MOMENT OF FORCE

The turning effect referred to in this chapter is called the moment of a force and its magnitude depends on two quantities,

(1) the size of the *force* applied, and
(2) the *perpendicular distance* from the pivot or axis to the line of action of the force.

The moment of a force (M) is defined as *the product of the magnitude of force (F) and its perpendicular distance (d) from the pivot or axis to the line of action of the force* or

$$M = F \times d.$$

3.2.1. Units. From the definition of a moment of a force, the unit will be a combination of force and distance units. Thus if the force is in newtons and the distance in metres then from $F \times d = M$,

force (N) × perpendicular distance (m) = moment (N m)

The unit of the moment of a force is the newton-metre (N m).

EXAMPLE 3.1. In Figs 3.1, 3.2, and 3.3 determine the turning effect produced on each nut.

Case I (Fig. 3.1)

Turning effect on the nut = moment of the 30 N force about the nut.

Moment of force (M) = force (F) × perpendicular distance between geometrical centre of the nut and the line of action of the force (d),

or $M = F \times d$.

Moment = 30 N × 150 mm = 4·5 N m.

Turning effect of 30 N using 150 mm spanner = 4·5 N m.

Case II (Fig. 3.2)

Turning effect on the nut = moment of the 15 N force about the nut.

$M = F \times d$,

Moment = 15 N × 300 mm = 4·5 N m.

Turning effect of 15 N using 300 mm spanner = 4·5 N m.

Note: By doubling the length of spanner only half the hand force is required to produce the same turning effect on the nut.

Case III (Fig. 3.3)

Turning effect on the nut = moment of the 30 N force about the nut.

$M = F \times d$.

Note: In this case d is the perpendicular distance from the centre of the nut to the line of action of the force.

Using the sine ratio, $\dfrac{d}{150} = \sin 60°$,

$$d = 150 \times 0·866$$
$$= 129·9 \text{ mm}$$
$$\text{Say } d = 130 \text{ mm}$$

then moment = 30 N × 130 mm = 3·9 N m.

Turning effect of 30 N acting on 150 mm spanner at 60° to centre line of spanner = 3·9 N m.

Compare with Case I. The result of applying a force at 60° to the length of the spanner is to reduce the magnitude of the turning effect by approximately 14 per cent.

3.3. TERMS USED IN MOMENTS PROBLEMS

In problems on moments of a force the following terms are frequently used.

3.3.1. Fulcrum. The *fulcrum* is the point or axis about which rotation or turning takes place. In Example 3.1 the geometrical centre of the nut is considered to be the fulcrum.

3.3.2. Leverage or moment arm. The perpendicular distance from the line of action of the force to the fulcrum is known as the *leverage* or moment arm.

3.3.3. Clockwise moment. A *clockwise* moment tends to produce rotation of the body about the axis in a clockwise direction (↻). Clockwise moments are given a positive sign (+).

3.3.4. Anticlockwise moment. An *anticlockwise* moment tends to produce rotation of the body opposite to the clockwise moment (↺). Anticlockwise moments are given a negative sign (−).

3.3.5. Resulting moment. The *resulting* moment is the difference in magnitude between the total clockwise moment and total anticlockwise moment about the fulcrum.

EXAMPLE 3.2. Calculate the resulting moment about the fulcrum O for the force system shown in Fig. 3.4.

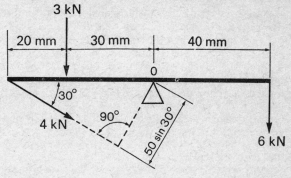

FIG. 3.4

SOLUTION

Moment = force × perpendicular distance from fulcrum O to line of action of the force.

Note 1: The perpendicular distance of the 4 kN force from the fulcrum equals 50 sin 30°.

The 6 kN force produces a clockwise moment about O.

Clockwise moment due to 6 kN force = 6 kN × 40 mm
= 240 N m.

The 3 kN force produces an anticlockwise moment about O.

Anticlockwise moment due to 3 kN force
= 3 kN × 30 mm
= 90 N m.

The 4 kN force produces an anticlockwise moment about O.

Anticlockwise moment due to 4 kN force
= 4 kN × (50 sin 30°) mm (see Note 1)
= 4 kN × (50 × ½) mm
= 4 kN × 25 mm
= 100 N m.

Total clockwise moment = 240 N m.
Total anticlockwise moment = (90+100) N m
$$= 190 \text{ N m.}$$
Resulting moment = clockwise moment—anticlockwise moment
$$= (240-190) \text{ N m}$$
Resulting moment = +50 N m, or 50 N m clockwise (see Note 2).

This result indicates that the system would tend to rotate in a clockwise direction.

Note 2: From the sign convention mentioned earlier in the chapter+50 N m indicates a clockwise moment.

3.4. PRINCIPLE OF MOMENTS

The *principle of moments* states that *when a body is in equilibrium under the action of a number of forces, the total clockwise moment about any point is equal to the total anticlockwise moment about the same point.*

3.5. LEVERS

A lever is a simple machine which applies the principle of moments to give mechanical advantage.

Levers may be straight or cranked, but the principle in each case is that a comparatively small force, usually termed *effort*, applied at a relatively large distance from the fulcrum, will either overcome or balance a greater force or *load* at a small distance from the fulcrum.

EXAMPLE 3.3. A uniform horizontal lever *AE* is supported on a fulcrum at *C* as shown in Fig. 3.5. Vertical

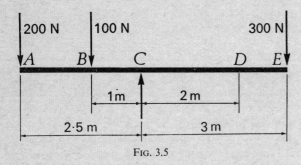

FIG. 3.5

forces of magnitudes 200 N, 100 N, and 300 N are applied to the lever as shown. What vertical force must be applied to the lever at *D* in order that the lever shall remain horizontal? Will the force at *D* act upward or downward? Neglect the weight of the lever.

U.L.C.I.

SOLUTION
Applying the principle of moments, for the lever to remain horizontal:

clockwise moments about *C* = anticlockwise moments about *C*.

Assume force at *D* acts downwards and has a magnitude *F*.

Taking moments about *C*,
total clockwise moments = $(F \times 2 + 300 \times 3)$ N m
$$= (2F + 900) \text{ N m}$$
Total anticlockwise moments = $(100 \times 1 + 200 \times 2.5)$ N m
$$= (100 + 500) \text{ N m.}$$

For equilibrium,
$$2F + 900 = 100 + 500$$
$$2F = 600 - 900,$$
$$\therefore 2F = -300,$$
$$\text{and } F = \frac{-300}{82} = -150 \text{ N.}$$
$$\text{or } F = 150 \text{ N, upwards.}$$

The negative sign indicates that the wrong direction was initially assumed for the force at *D*. For the lever to remain horizontal a force of 150 N is required at *D* acting vertically upwards.

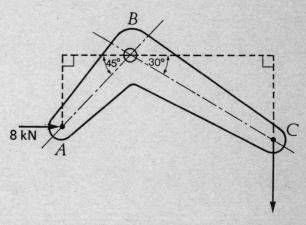

FIG. 3.6

EXAMPLE 3.4. Fig. 3.6 shows the cranked lever *ABC* pivoted at *B. AB* is 1 m and *BC* is 2 m. Calculate the magnitude of the vertical force at *C* required to balance a horizontal force of magnitude 8 kN applied at *A*.

SOLUTION
For the forces acting on the lever to balance:
clockwise moments about *B* must equal anticlockwise moments about *B*.

The force of magnitude 8 kN produces an anticlockwise moment about *B*.

Moment of 8 kN force about $B = (8 \times 1 \sin 45°)$ kN m
$$= (8 \times 1 \times 0.7071) \text{ kN m}$$
$$= 5.66 \text{ kN m.}$$

Let required vertical force at *C* be of magnitude *F*. This unknown force produces a clockwise moment about fulcrum *B*.

Moment of force of magnitude *F* about *B*

$$= F \times (2 \cos 30°)$$
$$= F \times (2 \times 0.866)$$
$$= (F \times 1.73)$$
$$= 1.73\,F$$

For equilibrium, applying principle of moments,

$$5.66 = 1.73\,F$$
$$\therefore F = \frac{5.66}{1.73} = 3.27 \text{ kN.}$$

Vertical force required at *C* to produce equilibrium equals 3·27 kN.

3.6. BEAMS

In the examples that follow a beam is considered to be a long straight bar carrying point loads at various positions along its length.

When considering problems involving reactions at beam supports and other beam loads the following points should be noted.

1. A uniform beam is one which has the same density of material and the same cross-sectional area throughout its length.
   The weight of the uniform beam may be considered to act at the centre of its length.
2. Where two unknown forces, including the reactions at the supports act on a beam, take moments about the position of one of the unknown forces.
   This procedure will prevent two unknown forces appearing in one equation.
3. For equilibrium of the beam, the total upward forces must equal the total downward forces.

EXAMPLE 3.5. A uniform beam *AB*, weighing 1 kN, is loaded as shown in Fig. 3.7. Calculate the reactions at the supports *A* and *B*.

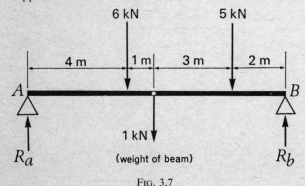

6 kN **5 kN**

4 m 1 m 3 m 2 m

A   *B*

1 kN ↓

(weight of beam)

R_a R_b

FIG. 3.7

SOLUTION

Let the reactions of the beam supports at *A* and *B* be R_a and R_b respectively.

Since *AB* is a uniform beam its weight of 1 kN may be

taken to act at the centre of the beam, i.e. 5 m from either end.

To obtain the reaction at *B*, take moments about *A*.

Total clockwise moment

$$= (6 \times 4 + 1 \times 5 + 5 \times 8) \text{ kN m}$$
$$= 24 \text{ kN m} + 5 \text{ kN m} + 40 \text{ kN m.}$$

Total clockwise moment $\quad = 69$ kN m.
Total anticlockwise moment $= (R_b \times 10)$ kN m
$$= 10 R_b \text{ kN m.}$$

Applying principle of moments,

$$10 R_b = 69$$
$$\therefore R_b = 6.9.$$

Reaction at support *B* = 6·9 kN.

To obtain the reaction at *A*, take moments about *B*.

Total clockwise moment

$$= (R_a \times 10) \text{ kN m}$$
$$= 10\,R_a \text{ kN m.}$$

Total anticlockwise moment

$$= (5 \times 2 + 1 \times 5 + 6 \times 6) \text{ kN m.}$$
$$= 10 \text{ kN m} + 5 \text{ kN m} + 36 \text{kN m.}$$

Total anticlockwise moment

$$= 51 \text{ kN m.}$$

Applying principle of moments,

$$10\,R_a = 51$$
$$\therefore R_a = 5.1.$$

Reaction at support   *A* = 5·1 kN.

Check: The total upward forces must equal the total downward forces.

Downward:
$$6 + 1 + 5 = 12.$$
Upward:
$$6.9 + 5.1 = 12.$$

EXAMPLE 3.6. Fig. 3.8 shows a uniform beam *AB*, simply supported at its ends and carrying vertical loads of 4 kN, 5 kN, and 7 kN at *C*, *D*, and *E* respectively. If the reaction at *B* is 13 kN determine

(a) the mass of the beam, given that $g = 9.81$ m/s², and
(b) the reaction at *A*.

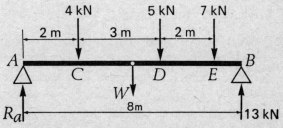

4 kN   **5 kN** **7 kN**

2 m 3 m 2 m

A   *B*

C   D   E

W↓

R_a 8m   13 kN

FIG. 3.8

SOLUTION

(a) The mass of the beam.
Let the weight of the beam be W acting at the centre of the uniform beam.

Since the reaction at A is not known, take moments about A.

Total clockwise moments about A

$$= (4000 \times 2 + W \times 4 + 5000 \times 5 + 7000 \times 7) \text{ N m}$$
$$= (8000 + 4W + 25\,000 + 49\,000) \text{ N m}$$
$$= (82\,000 + 4W) \text{ N m}.$$

Total anticlockwise moments about $A = (13\,000 \times 8)$
$$= 104\,000 \text{ N m}.$$

Applying principle of moments,

$$(82\,000 + 4W) = 104\,000,$$
$$4W = 104\,000 - 82\,000,$$
$$4W = 22\,000.$$
$$\therefore\ W = 5500.$$

Weight of beam = 5500 N.
Now, mass $m = W/g$,

$$\therefore\ \text{mass of beam} = \frac{5500}{9\cdot81} \text{ kg} = 560 \text{ kg}.$$

(b) The reaction at A.
The beam is in equilibrium, therefore total upward forces = total downward forces,
$$(13\,000 + R_a) = (4000 + 5000 + 7000 + 5500),$$
$$R_a = (21\,500 - 13\,000),$$
$$R_a = 8500.$$

Reaction at $A = 8\cdot5$ kN.

SUMMARY

The moment of a force is the turning effect of a force.

Moment of a force = force × moment arm length.
The unit is N × m = N m (newton-metre).

Principle of moments.
Clockwise moments = anticlockwise moments about the same point for a body in rotational equilibrium under the action of a number of forces.

Uniform beam.
The weight of a uniform beam may be taken to act at the centre of its length (weight = mg).

EXERCISES 3

1. What is the relationship which must exist between a system of clockwise and anticlockwise moments for them to be in equilibrium?

Most relevant section: **3.4**

U.L.C.I.

2. When pivoted at B, the lever shown in Fig. 3.9 requires a force of .... at point C to maintain equilibrium. If the pivot is moved 30 mm to the right, then is the required force at C increased or decreased? .... and by how much?

Most relevant sections: **3.4** and **3.5**

U.E.I.

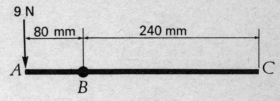

FIG. 3.9

3. State the 'principle of moments'.
A uniform horizontal beam 6 m long has a mass of 0·5 t and is supported at each end. The following masses are placed on the beam: 2 t at 1 m from the left-hand support and 4 t at 2 m from the right-hand support. Sketch the arrangement and calculate the reactions at the supports. Take g as 9·81 m/s².

Most relevant sections: **3.4** and **3.6**

N.C.T.E.C.

4. Fig. 3.10 shows part of a small clamping device. The arms of the bell-crank lever are at 90° and measure 160 mm and 75 mm, respectively, between centres, the longer arm being vertically upright. Find the intensity of pressure acting on the bottom of the pad of diameter 20 mm, when a force of magnitude 240 N is exerted at right-angles to the end of the longer arm.

Most relevant sections: **3.4, 3.5**, and **1.5.2**

U.E.I.

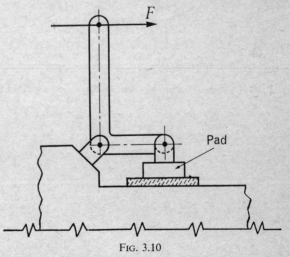

FIG. 3.10

5. Fig. 3.11 shows a tool clamped in the holder of a lathe by a force *F* of magnitude 900 N. Find the vertical force on (i) the tool and (ii) the packing.

Most relevant section: 3.4

C.G.L.I.

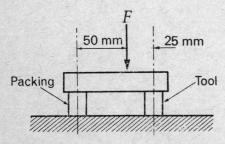

FIG. 3.11

6. A uniform beam is 4 m long and is simply supported at its ends. If the beam weighs 300 N determine the maximum distance from the left hand support of a load of 500 N if the value of the left-hand reaction is 600 N.

Most relevant section: 3.6

U.E.I.

7. A bell-crank lever is shown in Fig. 3.12, carrying a force of magnitude 35 N at *B*.
Calculate (a) the magnitude of a force *P* which is required to be applied at *A* in order to balance the lever.
(b) Either graphically or by calculation determine the magnitude and direction of the reaction at *O*.

Most relevant sections: 3.4 and 2.3

U.L.C.I.

8. A uniform beam *AB* is 3 m long and has a mass of 70 kg. It is simply supported at the left-hand end *A*, and at a point *C* which is 2·5 m to the right of *A*. Masses of 400 kg, 300 kg, and 200 kg are carried at points 0·6 m, 1·8 m, and 3 m to the right of *A* respectively. Find the reaction at each support.

Most relevant section: 3.6

C.G.L.I.

9. A horizontal beam *AB* is 6 m long and is freely supported at *A* and *B*. Loads of 30 kN and 60 kN are carried at points 2 m and 4 m from *A* respectively.

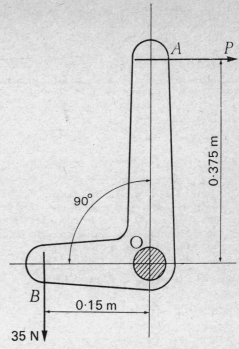

FIG. 3.12

(a) Find the magnitude of the reactions at *A* and *B*.
(b) If an extra load of 50 kN is to be added to the beam in such a position that the reactions at *A* and *B* are to be equal, what will these reactions then be, and at what distance from *A* must the 5 kN load be situated?

Most relevant section: 3.6

U.E.I.

10. In a laboratory experiment on the reactions at the supports of a beam, the beam was suspended from spring balances at *A* and *B* and loads applied at *C* and *D* as shown in Fig. 3.13.

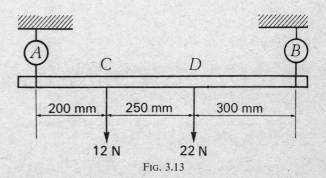

FIG. 3.13

(a) Calculate the theoretical reactions at A and B caused by the loads at C and D.
(b) The spring balances at A and B read 18·8 N and 17·7 N respectively.
Calculate the percentage error in each reading.
(c) Suggest one cause for the large error and explain carefully how you would correct for this.

Most relevant section: 3.6

C.G.L.I.

11. State the principle of moments.
A wagon has axles 4 m apart. A load of 70 kN is placed on the centre line between the axles, 0·75 m from the front axle. At each axle the load is supported by a spring and the stiffness of each spring is 2 kN/mm. Find the deflection of each spring.

Most relevant sections: 3.4, 3.6, and 5.4.1

E.M.E.U.

12. (a) State the principle of moments.
(b) Determine the reactions R_1, R_2 at the supports of the simply supported beam shown in Fig. 3.14.

Most relevant sections: 3.4 and 3.6

U.E.I.

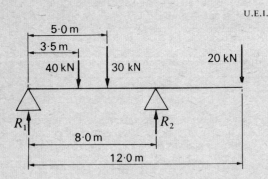

FIG. 3.14

13. In the lever-operated safety valve shown in Fig. 3.15 the pressure on the piston is not to exceed 400 kPa. If the area of the underside of the valve is 4 cm² determine the value of the force F which will prevent the valve from opening until the specified pressure is reached.

Most relevant sections: 1.5 and 3.5

U.E.I.

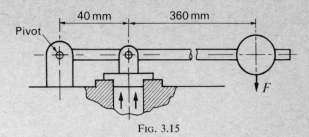

FIG. 3.15

14. (a) For the beam loaded as shown in Fig. 3.16, obtain expressions for the reactions at A and at B.
(b) Determine the value of W so that one reaction will be twice that of the other.

Most relevant section: 3.6

C.G.L.I.

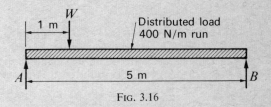

FIG. 3.16

15. (a) State the principle of moments.
(b) Fig. 3.17 shows a steel shaft 1 m long, of weight 450 N, supported in bearings at A and B distance 800 mm apart. In addition to the weight of the shaft, which is considered to act at its mid-point, there are two additional loads of 200 N and 350 N positioned as shown. Calculate
  (i) the reaction forces at A and B; and
  (ii) the position to which bearing B must be moved measured from the right-hand end of the shaft, so that both bearings have equal reactions (i.e. each support half of the total load).

Most relevant sections: 3.4 and 3.6

U.E.I.

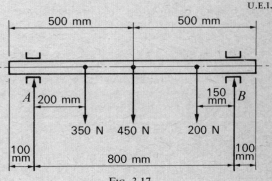

FIG. 3.17

ANSWERS TO EXERCISES 3

2. 3 N, increased, 1·71 N. **3.** 31·9 kN each. **4.** 1·63 MPa.
5. (i) 600 N, (ii) 300 N. **6.** 0·4 m. **7.** (a) 14 N, (b) 37·7 N,
21° 48′ upwards to left of vertical. **8.** $A = 3·69$ kN,
$C = 5·83$ kN. **9.** (a) $A = 40$ kN, $B = 50$ kN, (b) 70 kN
each, 2·4 m. **10.** (a) $A = 17·6$ N, $B = 16·4$ N, (b)
$A = 7·14$ per cent, $B = 7·9$ per cent. **11.** 28·4 mm front,
6·56 mm rear. **12.** (b) 23·75 kN, 66·25 kN. **13.** 16 N.
14. (a) $R_A = (0·8W+1000)$ N, $R_B = (0·2W+1000)$ N,
(b) 2·5. **15.** (b) (i) $R_A = 525$ N, $R_B = 475$ N, (ii) 140 mm.

4 Centre of gravity

4.1. CENTRE OF GRAVITY

In the previous chapter a number of problems involved uniform beams and in each of these problems the total weight of the beam was assumed to act at the centre of the beam. This assumption is now given further consideration.

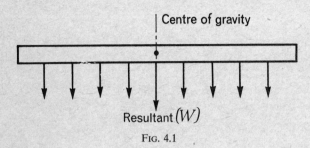

Centre of gravity

Resultant (W)

FIG. 4.1

The action of the earth's pull on every particle of the beam (Fig. 4.1) is represented by a large number of small forces or weights distributed over the entire length of the beam. These small forces are directed towards the centre of the earth and for all practical purposes their lines of action are assumed to be parallel.

The total weight W of the beam is the *resultant* of all these small parallel forces and acts in the same direction. The point at which this resultant W acts is called the *centre of gravity* of the beam. Although a uniform beam has been considered the reasoning given applies to a solid or volume of any shape.

The centre of gravity (c.g.) of a body is the point through which the resultant weight of the body acts. Table 4.1 gives the position of the centre of gravity of a number of common volumes.

4.2. SLINGING AND HOISTING

A feature of the centre of gravity is that any freely suspended body will hang so that its centre of gravity is vertically below the point of suspension.

This characteristic is important when a machine or heavy load is to be moved by slinging. It is essential for safe handling that the slings be of such length that the hook of the lifting tackle is positioned directly above the centre of gravity of the load to be moved. If the centre of gravity of the load is not in alignment with the hook of the lifting tackle an unbalanced moment is introduced which will cause the load to cant from the horizontal position (see Fig. 4.2).

TABLE 4.1

Shape of volume	Position of c.g. distant x from end shown	Volume
Cylinder	$h/2$	$\pi r^2 h$
Sphere	r	$4\pi r^3/3$
Hemisphere	$3r/8$	$2\pi r^3/3$
Cone	$h/4$	$\pi r^2 h/3$
Pyramid	$h/4$	$abh/3$

4.3. STABILITY

When a body is placed on a horizontal or an inclined surface, it will be in stable equilibrium provided that a vertical line through the centre of gravity is within the base of the body.

The objects shown in Figs 4.3(a) and 4.4(b) are stable and will remain at rest, while those shown in Figs 4.3(b) and 4.4(a) are unstable and will topple over as indicated.

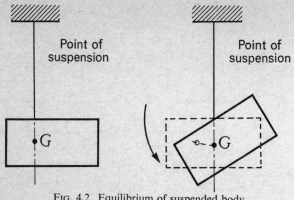

FIG. 4.2. Equilibrium of suspended body

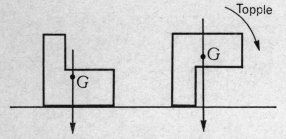

FIG. 4.3. (a) Stable; (b) Unstable

4.4. CENTRE OF GRAVITY OF A COMPO-
SITE SOLID

The position of the centre of gravity of a composite body is found by using a variation of the principle of moments;

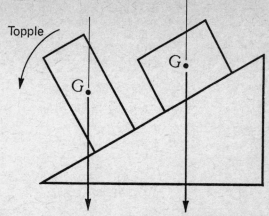

FIG. 4.4. (a) Unstable; (b) Stable

i.e. the moment of the resultant about a point is equal to the sum of the moments of its parts about the same point. The following worked examples illustrate the application of this principle to find the centres of gravity of composite bodies.

EXAMPLE 4.1. Fig. 4.5 shows a large shaft composed of two parts A and B. Part A is 2·5 m long and weighs 32 kN and part B is 2·5 m long and weighs 14 kN. Calculate the position of the centre of gravity of the shaft.

SOLUTION

G_1 is the known position of the centre of gravity of part A. G_2 is the known position of the centre of gravity of part B.

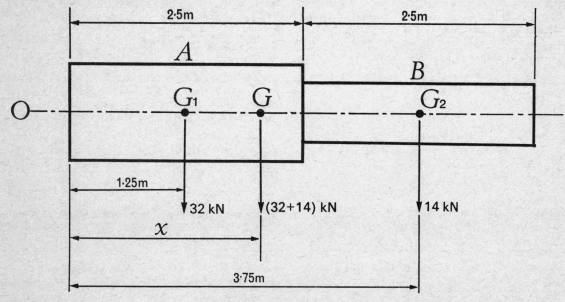

FIG. 4.5

Let the centre of gravity of the complete shaft be at the point G a distance x from the left-hand end.

To find x, apply the principle stated earlier in the chapter and take moments about the end O.

Moment of part A about O + moment of part B about O = moment of $(A+B)$ about O,

$(32 \text{ kN} \times 1\cdot25 \text{ m}) + (14 \text{ kN} \times 3\cdot75 \text{ m})$

$$= (32+14) \text{ kN} \times x \text{ m},$$
$$40 \text{ kN m} + 52\cdot5 \text{ kN m} = 46\, x \text{ kN m},$$
$$92\cdot5 \text{ kN m} = 46\, x \text{ kN m},$$
$$\therefore \ x = \frac{92\cdot5}{46},$$
$$= 2\cdot01 \text{ metres.}$$

The distance of the centre of gravity of the shaft from the left-hand end is $2\cdot01$ m.

EXAMPLE 4.2. An aluminium cube of 100 mm side is attached to a brass cube of 100 mm side as shown in Fig. 4.6. Calculate the position of the c.g. of the composite body. Take the densities of aluminium and brass as $2\cdot7$ g/cm³ and $8\cdot5$ g/cm³ respectively, and the acceleration due to gravity g as 10 m/s².

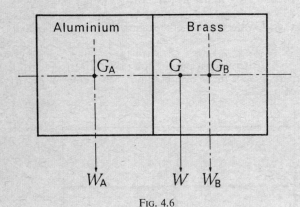

FIG. 4.6

SOLUTION

$$\text{Volume of each cube} = 100 \times 100 \times 100 \text{ mm}^3$$
$$= 10^6 \text{ mm}^3$$
$$= \frac{10^6}{10^3} \text{ cm}^3 = 1000 \text{ cm}^3.$$

Mass of aluminium cube, $m_A = \text{volume} \times \text{density}$
$$= 1000 \times 2\cdot7$$
$$= 2\cdot7 \text{ kg.}$$

Weight of aluminium cube, $W_A = m_A g$
$$= 2\cdot7 \times 10 \text{ N}$$
$$= 27 \text{ N.}$$

Similarly, weight of brass cube, $W_B = 85$ N.

Let the known positions of the c.g. of the separate sections be G_A and G_B as shown.

Distance between G_A and G_B = 100 mm or $0\cdot1$ m.

Let c.g. of composite body be at G, which is x m to the right of G_A or $(0\cdot1-x)$ m to the left of G_B.

Then taking moments about G,

$$\text{clockwise moment} = W_B \times (0\cdot1-x) \text{ m}$$
$$= 85 \times (0\cdot1-x) \text{ m}$$
$$= 8\cdot5 - 85x \text{ N m};$$
$$\text{anticlockwise moment} = W_A \times x \text{ m}$$
$$= 27x \text{ N m.}$$

Applying principle of moments,
$$27x = 8\cdot5 - 85x,$$
$$112x = 8\cdot5,$$
$$x = \frac{8\cdot5}{112} = 0\cdot076 \text{ m or 76 mm.}$$

Centre of gravity of composite body is $0\cdot076$ m (= 76 mm) to the right of G_A, i.e. 24 mm from the interface and in the brass cube.

Note: From Chapter 1, it will be remembered that mass $\propto$ density. This problem could, therefore, have been solved by using densities instead of weights as the volumes were equal.

4.5. CENTROID OF AN AREA

The term *centroid* is applied to areas in much the same way that centre of gravity is applied to solids. *The centroid of an area is the point at which the whole area may be considered to be concentrated.*

Note: The centre of gravity of a solid is the point at which the *whole weight* of a solid may be considered to be concentrated.

The following table gives the position of the centroid of a number of regular shaped areas. The symbols x and y represent the distances of the centroid from the left-hand and lower extremities of the area respectively.

4.5.1. Centroid of a triangle. The position of the centroid of any triangle is given by each of the following two methods.

1. In Fig. 4.7 the line joining the mid-point of a side of the triangle to the opposite angle is called a 'median'. The centroid of the triangle is at the point where the medians intersect.

Note: It is necessary to draw only two of the medians.

2. In Fig. 4.8 the centroid of the triangle is shown at the point measured one-third of the *perpendicular* distance from any side to the opposite angle.

TABLE 4.2

Shape of area	Distance x	Distance y	Area
Square	$\dfrac{a}{2}$	$\dfrac{a}{2}$	a^2
Rectangle	$\dfrac{a}{2}$	$\dfrac{b}{2}$	ab
Circle	r	r	πr^2
Semi-circle	$\dfrac{4r}{3\pi}$	r	$\dfrac{\pi r^2}{2}$
Right-angled triangle	$\dfrac{b}{3}$	$\dfrac{h}{3}$	$\dfrac{bh}{2}$

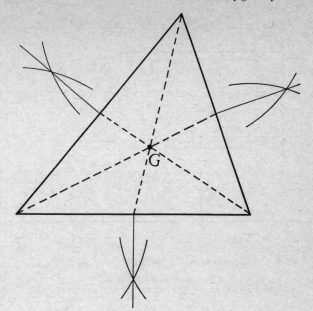

FIG. 4.7. Location of centroid of triangle (1)

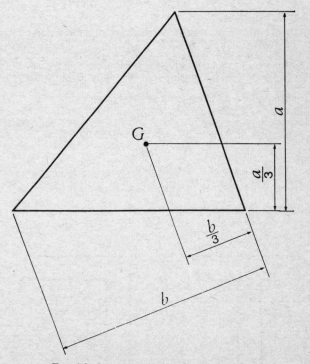

FIG. 4.8. Location of centroid of triangle (2)

4.5.2. Centroid of a composite area. The method adopted is similar to that applied in finding the centre of gravity of a composite solid.

The area is divided into simple parts such as triangles, rectangles, and quadrants of circles.

The principle is then applied that the moment of the whole area about an axis is equal to the sum of the moments of the parts about the same axis.

Symbols x and y are used to indicate the horizontal and vertical distances respectively of the centroid from some datum line or point.

EXAMPLE 4.3. Find the position of the centroid of the composite area shown in Fig. 4.9.

SOLUTION

The composite area consists of a square A and triangle B.

Let C_A be the known position of the centroid of the square and C_B be the known position of the centroid of the triangle.

Then $C_{(A+B)}$ represents the centroid position of the combined area.

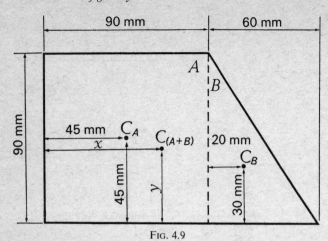

FIG. 4.9

Area of square A = 90 mm $\times$ 90 mm = 8100 mm^2

Area of triangle B = $\dfrac{60 \text{ mm} \times 90 \text{ mm}}{2}$ = 2700 mm^2.

Area of combined figure $(A+B)$ = 8100 mm^2+2700 mm^2
$$= 10\ 800 \text{ mm}^2.$$

To find distance x,
take moments about the left-hand edge of the square
and apply the principle.

Moment of area of square A+moment of area of
triangle B = moment of area of (square A+triangle B),

8100 mm$^2 \times$ 45 mm+2700 mm$^2 \times$ 110 mm
$$= 10\ 800\ x \text{ mm}^3,$$
$$364\ 500 \text{ mm}^3+297\ 000 \text{ mm}^3 = 10\ 800\ x \text{ mm}^3,$$
$$x = \frac{661\ 500 \text{ mm}^3}{10\ 800 \text{ mm}^2} = 61 \text{ mm}$$

Distance of centroid from left-hand edge = 61 mm.

To find distance y,
take moments about the lower edge of the square and
triangle and apply the principle.

Moment of area A+moment of area B

$$= \text{moment of area } (A+B),$$
$$8100 \text{ mm}^2 \times 45 \text{ mm} + 2700 \text{ mm}^2 \times 30 \text{ mm}$$
$$= 10\ 800y \text{ mm}^3,$$
$$364\ 500 \text{ mm}^3 + 81\ 000 \text{ mm}^3 = 10\ 800y \text{ mm}^3,$$
$$445\ 500 \text{ mm}^3 = 10\ 800y \text{ mm}^3,$$
$$y = \frac{445\ 500 \text{ mm}^3}{10\ 800 \text{ mm}^2} = 41 \text{ mm}.$$

Distance of centroid from lower edge of area = 41 mm.

Note: As would be expected the centroid of the composite area lies between the centroid of the square and the centroid of the triangle and also lies on the line joining these two points.

4.5.3. Beam sections. It is important for purposes of design to be able to locate the centroids of beam sections. The cross-section of a beam is designed in such a way as to provide both efficient and economic use of material.
Fig. 4.10 shows various standard structural sections.

EXAMPLE 4.4. Calculate the position of the centroid of the beam section shown in Fig. 4.11.

SOLUTION

The section consists of rectangular flanges A and C and a rectangular web B.
Let the known positions of the centroids of the three rectangles be C_A, C_B, and C_C, then $C_{(A+B+C)}$ represents the centroid of the whole section.
Since the section is symmetrical about a vertical axis passing through the centre of the web, only the distance y mm of the centroid from the lower edge requires calculating.

Area of top flange A = 150 $\times$ 30=4500 mm^2.
  Area of web B = 200 $\times$ 50=10 000 mm^2.
  Area of flange C = 250 $\times$ 50 = 12 500 mm^2.
Area of section
$$(A+B+C) = 4500+10\ 000+12\ 500 = 27\ 000 \text{ mm}^2.$$

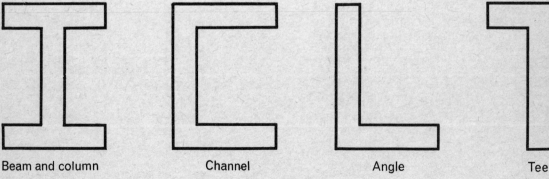

Beam and column   Channel   Angle   Tee

FIG. 4.10. Beam sections

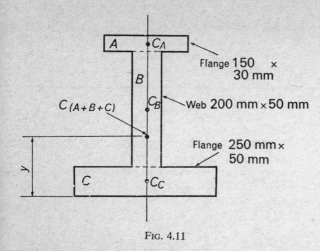

FIG. 4.11

In this case the mass of the lamina is distributed in the same way as its area. It is therefore more convenient to refer to the centroid of a lamina than to its centre of gravity.

These points do not coincide since the centre of gravity of a lamina would be at the middle of its thickness.

EXAMPLE 4.5. A thin circular steel plate of diameter 140 mm has a hole of diameter 40 mm punched in it. The distance between the centres of the plate and punched hole is 30 mm.

Calculate the position of the centroid of the plate shown in Fig. 4.12 (where diameter is indicated by the symbol $\varnothing$).

Take moments of the areas about the lower edge of flange C and use the principle: moment of area of flange A+moment of area of web B+moment of area of flange C = moment of area of section $(A+B+C)$.

$4500 \text{ mm}^2 \times 265 \text{ mm}+10\,000 \text{ mm}^2 \times 150 \text{ mm}$
$+12\,500 \text{ mm}^2 \times 25 \text{ mm} = 27\,000 \text{ mm}^2 \times y \text{ mm},$
$(1\,192\,500+1\,500\,000+312\,500) \text{ mm}^3 = 27\,000\,y \text{ mm}^3,$

$$y = \frac{3\,005\,000}{27\,000} = 111.$$

The centroid of the section lies on the axis of symmetry and is situated approximately 111 mm from the lower edge of the bottom flange.

4.5.4. Centroid of a plane lamina. A lamina is a thin plate or sheet of material having uniform thickness and density.

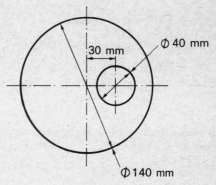

FIG. 4.12

SOLUTION

In this case the centroid lies on the axis of symmetry which is the line joining the centre of the plate and the centre of the punched hole in Fig. 4.13.

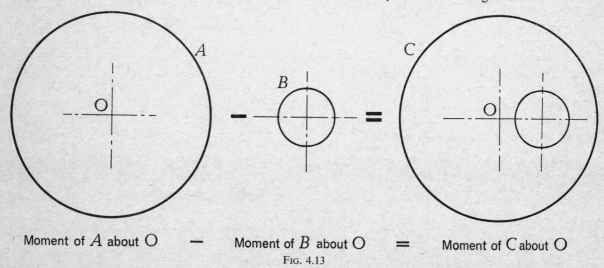

Moment of A about O — Moment of B about O = Moment of C about O

FIG. 4.13

To simplify calculations take moments of areas of A, B, and C about the centre of the plate.

$$\text{Area of unpierced plate} = \frac{\pi \times 140^2}{4} \text{ mm}^2.$$

$$\text{Area of punched hole} = \frac{\pi \times 40^2}{4} \text{ mm}^2.$$

$$\text{Area of pierced plate} = \frac{\pi \times 140^2}{4} - \frac{\pi \times 40^2}{4} \text{ mm}^2.$$

Let x be the distance of the centroid of the pierced plate from the centre of the plate.

Take moments of the areas about the centre of plate and apply the principle,

moment of area of unpierced plate−moment of area of punched hole = moment of area of pierced plate,

$$\pi \times 70^2 \times 0 - \pi \times 20^2 \times 30 = \pi(70^2 - 20^2)x,$$

$$0 - 12\,000 = 4500\,x,$$

$$x = \frac{-12\,000}{4500} = -2\cdot67 \text{ mm}.$$

The centroid lies on the line of symmetry at a distance of 2·67 mm to the left of the centre of the plate.

Note: The minus sign indicates that the centroid lies on the opposite side of the centre of the plate from the punched hole.

SUMMARY

The centre of gravity of a body is the point at which the whole weight of a body may be taken to act.

The centroid, often referred to as centre of area, is a point in an area or lamina where the whole area may be considered to be concentrated.

To find the centre of gravity of a composite solid: divide the solid into regular shapes, then apply the principle:

moment of solid about a given axis = sum of the moments of the parts about the same axis.

Note: When a solid has the same density throughout, volumes may be used instead of weights.

To find the centroid of composite areas:

divide the area into regular shapes, then apply the principle:

moment of total area about a given axis = sum of the moments of the parts about the same axis.

Slinging and hoisting.

It is important when lifting heavy loads or machines to ensure that the hoisting position is vertically above the centre of gravity of the load.

EXERCISES 4

1. A thin circular metal plate of uniform thickness has a diameter of 80 mm. A hole of diameter 40 mm is punched in the plate, the centre of the hole being 10 mm from the centre of the plate. Calculate the position of the centroid of the finished plate.

Most relevant section: 4.5.2

2. A roller has an overall length of 40 mm. A blind flat-bottomed hole in the roller has a depth of 24 mm, is concentric with the outside diameter, and has a diameter one-half that of the roller. Calculate the position of the centre of gravity of the roller, measured from the end containing the hole.

Most relevant section: 4.4

3. A uniform rod AB has a mass of 2 kg and has masses of 7 kg and 9 kg at its ends A and B as shown in Fig. 4.14. Find the position of the centre of gravity of the arrangement measured as a distance from end A.

Most relevant section: 3.4

U.L.C.I.

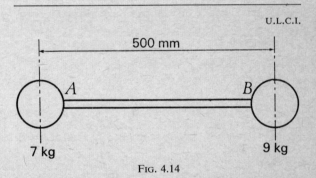

FIG. 4.14

4. Calculate the position of the centre of gravity of the solid object shown in Fig. 4.15.

Most relevant sections: 4.1, 4.5.2, and 4.5.4

U.E.I.

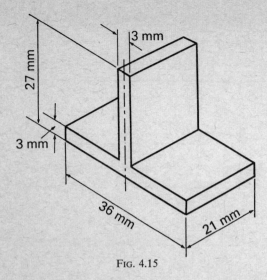

Fig. 4.15

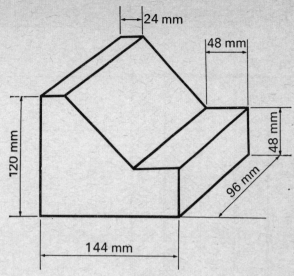

Fig. 4.17

5. Fig. 4.16 shows a steel plate 6 mm thick. (a) Calculate its mass, given that steel has a density of 7·75 g/cm³. (b) Find the position of the centre of gravity of the plate, expressing it as

   (i) its distance from *AB*,
   (ii) its distance from *BC*.

Most relevant sections: 1.3 and **4.5.2**

C.G.L.I.

7. Calculate the position of the centroid of the lamina shown in Fig. 4.18 measured from the 80 mm edge.

Most relevant section: 4.5.2

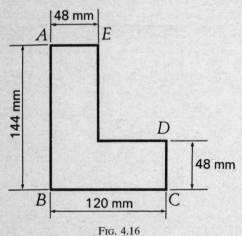

Fig. 4.16

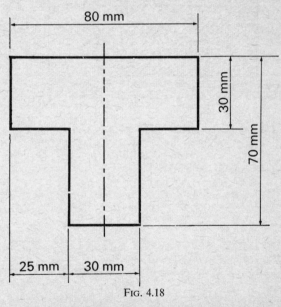

Fig. 4.18

6. Calculate the position of the centre of gravity of the object shown in Fig. 4.17.
How could your calculations be easily checked experimentally?

Most relevant sections: 4.1, 4.5.2, and **4.5.4**

U.E.I.

8. A cast steel block, having a hole of square section, is plugged with lead as shown in Fig. 4.19. Calculate the position of the c.g. measured from the top of the block. Take the densities of steel and lead as 7·75 g/cm³ and 11·35 g/cm³ respectively.

Most relevant sections: 4.4 and **1.3**

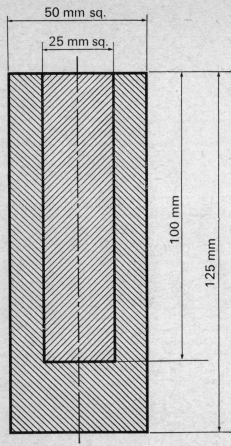

FIG. 4.19

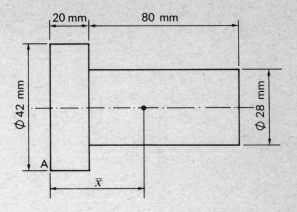

FIG. 4.20

9. A triangular metal plate, 30 mm × 40 mm × 50 mm and of uniform thickness, stands vertically on its smallest edge. Find, by scale drawing, the centroid of the side of the plate.

   If the plate is maintained in the vertical plane, what are the greatest angles in each direction to which the surface on which it rests may be tilted without the plate toppling over?

Most relevant sections: 4.5.1 and 4.3

10. (a) Define the term centre of gravity.
   (b) Calculate the position of the c.g, from face *A*, distance $\bar{x}$, of the steel pin shown in Fig. 4.20.

Most relevant sections: 4.1 and 4.4

U.E.I.

11. (a) State the Principle of moments.
   (b) Fig. 4.21 shows a template, with a 50 mm diameter hole drilled in it. Determine the position of its centroid, relative to the sides *AB* and *BC*.

Most relevant sections: 3.4 and 4.5.2

N.C.T.E.C.

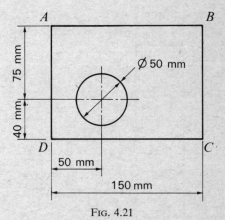

FIG. 4.21

12. Calculate the position of the c.g. for the 6 mm thick metal plate shown in Fig. 4.22. The hole is in the form of a right-angled triangle.

Most relevant sections: 4.1 and 4.5.2

N.C.T.E.C.

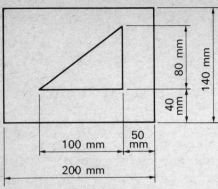

FIG. 4.22

13. A horizontal stepped spindle having an overall length of 500 mm has its diameter reduced progressively as follows: the spindle is 150 mm diameter for a length of 200 mm, 100 mm diameter for a further 150 mm, and 50 mm diameter for the remaining 150 mm.

Calculate:

(a) The mass of the spindle if the density of the material is 7.7×10^3 kg/m³.
(b) The distance of the centre of gravity of the spindle from the larger end.

Most relevant sections: 1.3.1 and **4.4**

Y.C.F.E.

14. A vehicle, together with its load, weighs 8·5 kN. If the distance apart of the wheels is 2·5 m and the centre of gravity of vehicle and load is central and 3 m above the road, calculate the maximum angle to which the vehicle can tilt and not fall over.

Most relevant section: 4.3

15. Find the position of the centre of gravity of the casting shown in Fig. 4.23 (diameter indicated by $\varnothing$). The casting is symmetrical about the axis *AB*.

Most relevant section: 4.4

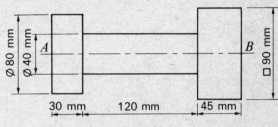

FIG. 4.23

ANSWERS TO EXERCISES 4

1. 3·33 mm from centre of plate, away from hole. **2.** 21·4 mm. **3.** 277·7 mm. **4.** 6·9 mm up from base, on centre line; 10·5 mm from front or back. **5. (a)** 0·482 kg, **(b) (i)** 43·92 mm, **(ii)** 55·92 mm. **6.** 57·12 mm from left hand edge; 44·4 mm from bottom edge; 48 mm from front or back. **7.** 26·7 mm. **8.** 61·4 mm. **9.** 10 mm from 40 mm edge; 13·3 mm from 30 mm edge. 45°, 63·5°. **10.** 42 mm. **11.** 71·8 mm from *BC*; 55·3 mm from *AB*. **12.** 97·2 mm from *LH* edge; 70·6 mm from bottom edge. **13. (a)** 38·5 kg, **(b)** 160 mm. **14.** 22° 37′. **15.** 118 mm from *A*.

5 Force and materials

5.1. APPLICATION OF FORCES TO SOLID MATERIALS

The various effects of the application of force to a solid material depend upon the magnitude and type of force, and the mechanical properties of the material.

Fig. 5.1 shows part of a structure, and illustrates the three fundamental effects that forces have on materials.

The load F is transmitted through the plate P, bolt B, and angle-irons A, to the fixed supports S. The plate is subjected to a tensile force tending to pull the plate apart; the bolt is subjected to shearing forces trying to cut across the bolt; the angle-irons and supports are subjected to compressive forces tending to crush the materials.

5.2. MECHANICAL PROPERTIES OF MATERIALS

When a material is selected for service, the designer estimates the probable loads to be carried by the part being designed, and with a knowledge of the properties of the material to be used is able to determine the shape and size required.

5.2.1. Strength. The *strength* of a material is its capacity to resist force without fracture; the force may be tensile, compressive, or shear.

The term *tenacity* is often used to refer to strength in tension.

5.2.2. Elasticity. *Elasticity* is the property by which a material returns exactly to its original shape and size on the removal of a force.

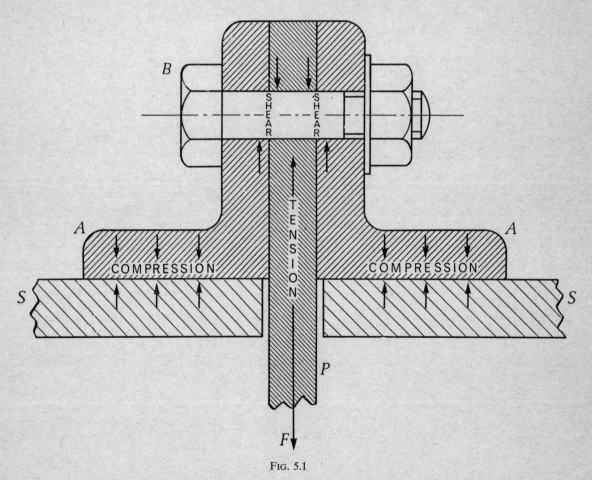

FIG. 5.1

5.2.3. Plasticity. A *plastic* material will retain its deformed shape after the removal of an applied force.

5.2.4. Ductility. The property which enables a material to be drawn out in length (e.g. wire drawing) is known as *ductility*.

5.2.5. Malleability. A *malleable* material has the capacity to be deformed in all directions; the material may be hammered or worked into the required shape (e.g. forging, stamping, hot rolling, etc.).

5.2.6. Hardness. *Hardness* is generally considered as the resistance of a material to indentation, scratching, and wear. In general, a hard material possesses little ductility.

5.2.7. Brittleness. *Brittleness* causes a material to break without warning and with little elongation under load. It may be considered the reverse of ductility.

5.2.8. Toughness. The resistance to impact and shock loads, and to repeated bending and twisting, is known as the *toughness* of a material.

5.3. STRESS AND STRAIN

Consider a bar of material which is subjected to an external force. A resisting force is set up within the bar, and the material is said to be in a state of stress.

There are three fundamental types of stress to be considered:

1. *Tensile stress* which is set up by forces tending to pull the material apart.
2. *Compressive stress* produced by forces tending to crush the material.
3. *Shear stress* resulting from forces tending to cut through the material, i.e. tending to make one part of the material slide across the other part.

Intensity of stress is a measure of the load carried by unit area of the material.

Let F = applied force,
  A = area of material resisting the force,
  σ = direct stress (i.e. tensile and compressive stress),
  τ = shear stress,

Then

$$\text{direct stress } \sigma = \frac{\text{normal force } F}{\text{area } A}$$

and

$$\text{shear stress } \tau = \frac{\text{shear force } F}{\text{area } A}.$$

The special name pascal (Pa) has been approved for the SI unit of stress. Derived units of stress are N/m^2, N/mm^2, and MN/m^2, where $1\,N/m^2 = 1\,Pa$. The meganewton per square metre (MN/m^2) is identical with the newton per square millimetre (N/mm^2).

For tensile and compressive stresses (Figs 5.2(a) and (b)) the area to be considered is at right angles to the line of action of the applied force, while for shear stress (Fig. 5.2(c)) the area lies in the same direction as the applied force.

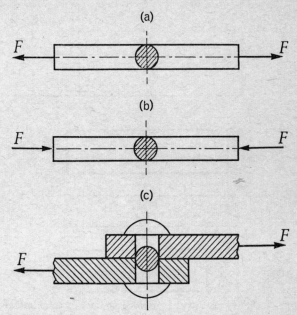

Fig. 5.2(a) Tensile stress; (b) compressive stress; (c) shear stress

The external force acting on the piece of material will cause a change in the shape of the material; e.g., a tensile force will cause it to stretch.

A piece of material which is altered in shape due to a force is said to be strained. Strain is the ratio of the deformation produced to the original dimension, and since both quantities are measured in the same units, strain is dimensionless, i.e. it has no units.

Let x = deformation,
  l = original dimension,
  ε = direct strain,
  γ = shear strain.

Then

$$\text{direct strain } \varepsilon = \frac{\text{deformation } x}{\text{original dimension } l}.$$

Likewise

$$\text{shear strain } \gamma = \frac{x}{l}.$$

Note: For shear strain the deformation is measured perpendicular to the original dimension (Fig. 5.3(a)) while for direct strains the distortion is measured in line with the original dimension (Figs 5.3(b) and (c)).

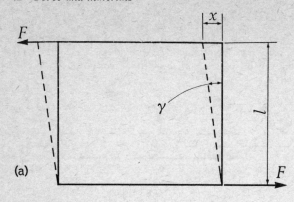

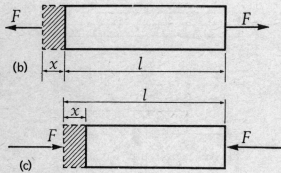

(a)

(b)

(c)

FIG. 5.3(a) Shear strain; (b) tensile strain; (c) compressive strain

EXAMPLE 5.1. A 20 mm diameter steel rod, 1 m long, carries a load of 45 kN which causes an extension of 1·8 mm. Calculate the stress and strain in the rod.

U.E.I.

SOLUTION

$$\text{Stress } \sigma = \frac{\text{normal force } F}{\text{area } A}$$

$$= \frac{45 \text{ kN}}{\pi (10)^2 \text{ mm}^2}$$

$$= \frac{450}{\pi} \text{ N/mm}^2$$

$$= 143 \text{ N/mm}^2$$

$$= 143 \text{ MN/m}^2 \text{ or MPa.}$$

$$\text{Strain } \varepsilon = \frac{\text{extension } x}{\text{original length } l}$$

$$= \frac{1·8 \text{ mm}}{1 \times 1000 \text{ mm}}$$

$$= 0·001\,8.$$

EXAMPLE 5.2. The minimum force required to punch a rectangular hole 35 mm × 25 mm through a metal strip 3 mm thick is 140 kN. Calculate the shear stress in the material.

SOLUTION

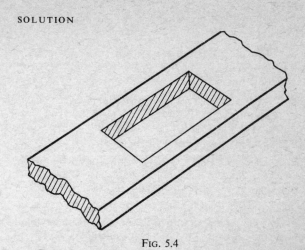

FIG. 5.4

Fig. 5.4 shows the hole in the strip, and clearly indicates the 'area' which is to be considered.

Area to be sheared

$$= \text{perimeter of hole} \times \text{thickness of strip}$$

$$= (35 + 35 + 25 + 25) \times 3 \text{ mm}^2$$

$$= 360 \text{ mm}^2.$$

$$\text{Shear stress } \tau = \frac{\text{force } F}{\text{area } A}$$

$$= \frac{140 \text{ kN}}{360 \text{ mm}^2}$$

$$= 389 \text{ N/mm}^2$$

$$= 389 \text{ MN/m}^2 \text{ or MPa.}$$

EXAMPLE 5.3. A short metal column having the cross-section shown in Fig. 5.5 is to carry an axial compressive load. If the compressive stress in the material is not to exceed 80 N/mm² or MPa, calculate the maximum load that may be placed on the column.

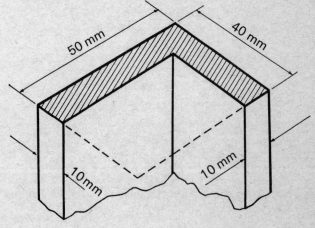

FIG. 5.5

SOLUTION

$$\text{Area of cross-section of column} = \frac{\text{large}}{\text{rectangle}} - \frac{\text{small}}{\text{rectangle}}$$

$$= (50 \times 40) - (40 \times 30) \text{ mm}^2$$

$$= 2000 - 1200 \text{ mm}^2$$

$$= 800 \text{ mm}^2.$$

$$\text{Stress } \sigma = \frac{\text{normal force } F}{\text{area } A},$$

so that force $F = A\sigma$

$$= 800 \times 80 \text{ kN}$$

$$= 64 \text{ kN}.$$

5.4. HOOKE'S LAW

This important law states that *within the elastic range of a material the deformation is directly proportional to the applied force producing it.*

5.4.1. Spring stiffness. Hooke's law also applies to springs as will be readily understood from a consideration of a spring balance which is used for measuring weight.

The stiffness of a spring is the force required to cause unit deflection.

$$\text{Stiffness} = \frac{\text{force}}{\text{deflection}}.$$

EXAMPLE 5.4. A helical spring shortens by 25 mm when a compressive load of 150 N is applied. Calculate the total shortening if an additional 90 N were applied.

SOLUTION

$$\text{Spring stiffness} = \frac{\text{force}}{\text{deflection}}$$

$$= \frac{150 \text{ N}}{25 \text{ mm}} = 6 \text{ N/mm}$$

(thus every 6 N causes a shortening of 1 mm).
Total load applied to spring $= (150 + 90) \text{ N} = 240 \text{ N}$,

$$\therefore \text{ total shortening} = \frac{240 \text{ N}}{6 \text{ N/mm}} = 40 \text{ mm}.$$

5.4.2. Modulus of elasticity. From a consideration of Hooke's law, it follows that stress is directly proportional to strain while the material remains elastic. Then

$$\text{stress} = \text{strain} \times \text{a constant}$$

so that $\quad \dfrac{\text{stress}}{\text{strain}} = \text{a constant}.$

The constant (symbol E) is known as the *modulus of elasticity (or Young's modulus)* and is measured in the same units as stress.

$$E = \frac{\text{stress } \sigma}{\text{strain } \varepsilon}.$$

Now, since $\sigma = \dfrac{F}{A}$ and $\varepsilon = \dfrac{x}{l}$,

$$\text{then } E = \frac{F/A}{x/l},$$

$$\therefore E = \frac{Fl}{Ax}.$$

The value of E may be determined from the stress–strain (and load–extension) graph.

5.4.3. Modulus of rigidity. The relationship between shear stress and shear strain is known as the *modulus of rigidity* (symbol G).

$$G = \frac{\text{shear stress } \tau}{\text{shear strain } \gamma}.$$

The following table shows typical average values of the elastic constants for some common metals. These values may be affected by factors such as composition, manufacturing processes, etc.

TABLE 5.1

Material	Modulus of elasticity E (kN/mm² or GPa)	Modulus of rigidity G (kN/mm² or GPa)
Aluminium	70	26
Brass (70/30)	101	37
Cadmium	50	19
Chromium	279	115
Copper	130	48
Iron (cast)	152	60
Lead	16	6
Steel (mild)	212	82
Tin	50	18
Titanium	116	44
Tungsten	411	161
Zinc	108	43

EXAMPLE 5.5. A tie bar having a 25 mm square cross-section is 2 m long. When carrying a load of 33 kN its extension is 0·5 mm. Another tie bar of 20 mm square cross-section and of the same material has a length of 1·25 m, and carries a load of 25 kN. What will be the extension of this second tie bar?

SOLUTION

$$\text{Modulus of elasticity } E = \frac{\text{stress}}{\text{strain}} = \frac{Fl}{Ax}$$

$$= \frac{33 \text{ kN} \times (2 \times 1000) \text{ mm}}{(25)^2 \text{ mm}^2 \times 0·5 \text{ mm}}$$

$$= 211 \text{ kN/mm}^2 \text{ or GPa.}$$

Now, the value of E is also 211 kN/mm² or GPa for the second bar, since both bars are of the same material.

Then, from $E = \dfrac{Fl}{Ax}$,

extension of second bar $= \dfrac{Fl}{AE}$,

$$= \frac{25 \text{ kN} \times (1 \cdot 25 \times 1000) \text{ mm}}{(20)^2 \text{ mm}^2 \times 211 \text{ kN/mm}^2}$$

$$= 0 \cdot 37 \text{ mm}.$$

5.5. TENSILE STRENGTH

The *tensile strength* of a material is of importance to the designer when determining the allowable working stresses in the material under load.

The value of the tensile strength (or ultimate stress) for a material is determined from a tensile test to destruction.

$$\text{Tensile strength} = \frac{\text{maximum load carried during test}}{\text{original cross-sectional area of testpiece}}.$$

5.5.1. Factor of safety. In engineering it is generally important not to stress any material beyond its elastic limit: parts of machines and mechanisms must not be permanently deformed. To ensure that this does not occur, a maximum working stress is chosen within the elastic range of the material, and design calculations are based upon this figure.

$$\text{Factor of safety} = \frac{\text{tensile strength}}{\text{maximum allowable working stress}}$$

The maximum allowable working stress is generally a simple fraction of the tensile strength.

EXAMPLE 5.6. A truck of mass 5 t is being hauled up an inclined track at a uniform speed by a cable that runs parallel to the track. The inclination of the track is 20° to the horizontal.
(a) Find the tension in the cable, given $g = 10$ m/s².
(b) If the cable is made up of 96 strands of steel wire of 1·2 mm diameter, find the factor of safety for the cable given that the tensile strength of the steel is 600 MN/m² or MPa.

C.G.L.I.

SOLUTION

See Figs 5.6 and 5.7.

(*Note:* The tension in the cable may be found by either graphical or analytical means.)

$$\text{Mass of truck} = 5 \text{ t} = 5000 \text{ kg}.$$
$$\text{Weight of truck} = mg$$
$$= 5000 \times 10 \text{ kN}$$
$$= 50 \text{ kN}.$$
$$\text{Tension in cable} = 50 \times \sin 20° \text{ kN}$$
$$= 50 \times 0 \cdot 342 \text{ kN}$$
$$= 17 \cdot 1 \text{ kN}.$$

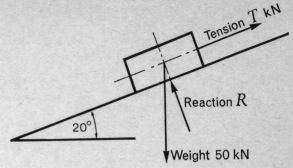

FIG. 5.6. Space diagram

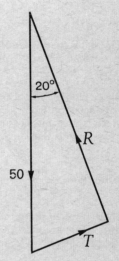

FIG. 5.7. Force (vector) diagram

Total cross-sectional area of 96 strands of steel wire of 1·2 mm diameter

$$= 96 \, (\pi r^2) \text{ mm}^2$$
$$= 96\pi \, (0 \cdot 6)^2 \text{ mm}^2$$
$$= 104 \text{ mm}^2.$$

Now, stress $= \dfrac{\text{force}}{\text{area}}$,

$$\therefore \text{ working stress} = \frac{17 \cdot 1 \text{ kN}}{104 \text{ mm}^2} = \frac{17 \, 100 \text{ N}}{104 \text{ mm}^2}$$

$$= 164 \text{ N/mm}^2 \text{ or MPa}.$$

$$\text{Factor of safety for cable} = \frac{\text{tensile strength}}{\text{working stress}}$$

$$= \frac{600}{164} \quad \begin{array}{l} (600 \text{ MN/m}^2 = \\ \quad\quad 600 \text{ N/mm}^2) \end{array}$$

$$= 3 \cdot 65.$$

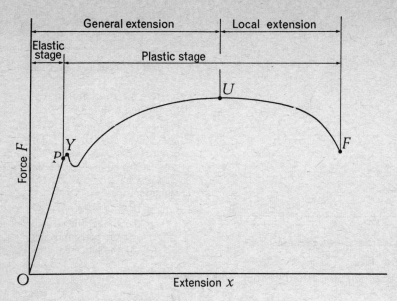

FIG. 5.8. Tensile test curve for mild steel

5.6. MECHANICAL TESTING OF MATERIALS

Mechanical tests are carried out to determine the properties of a material, to ascertain if the raw material meets the desired standard, and as a check upon manufacturing processes, for example, heat-treatment.

5.6.1. Tensile test. One of the principal mechanical tests carried out is the tensile test to destruction, in which the specimen of material to be tested is loaded in tension until fracture occurs.

Fig. 5.8 shows a typical load–extension curve for a specimen of mild steel, which is a ductile material. The point P, at the end of the straight line OP, is called the *limit of proportionality*. Between O and P the extension x is directly proportional to the applied force F; and the material obeys Hooke's law. The *elastic limit* is at, or very near to, the limit of proportionality. When the limit of proportionality has been passed, the extension ceases to be proportional to the load, and at the *yield point* Y, the extension suddenly increases, and the material enters the plastic stage. At point U the load is greatest: the tensile strength is determined from this point. The extension of the testpiece has been general up to point U, but at U 'waisting' occurs, and subsequent extension is local. Since the cross-sectional area of the testpiece is considerably reduced at the 'waist', the actual stress increases, resulting in a falling-off of the load, and fracture occurs at point F.

From the load–extension curve the following information can be determined:

$$\text{modulus of elasticity} = \text{slope of line } OP \times \left(\frac{\text{gauge length}}{\text{original cross-sectional area}}\right).$$

$$\text{yield stress} = \frac{\text{load at } Y}{\text{original cross-sectional area}}$$

$$\text{tensile strength} = \frac{\text{load at } U}{\text{original cross-sectional area}}$$

$$\text{breaking stress} = \frac{\text{load at } F}{\text{final cross-sectional area at fracture}}$$

$$\text{percentage elongation} = \left(\frac{\text{final gauge length} - \text{original gauge length}}{\text{original gauge length}}\right) \times \frac{100}{\text{per cent}}$$

$$\text{percentage reduction in area} = \left(\frac{\text{original area} - \text{final area at fracture}}{\text{original area}}\right) \times \frac{100}{\text{per cent.}}$$

Some typical load–extension curves for various common engineering materials are shown in Fig. 5.9.

An inspection of the curves shows that annealed copper is very ductile, while hard-drawn copper is stronger but less ductile. Hard-drawn 70/30 brass is seen to be both strong and ductile. The brittleness of cast iron is clearly indicated, and it is for this reason that cast-iron parts are rarely used when the load is tensile.

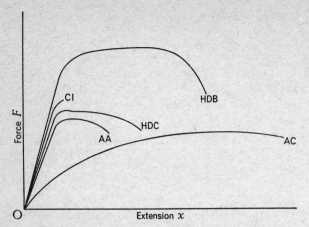

Fig. 5.9. Tensile test curves for various materials

Aluminium alloy is shown to be fairly strong and ductile.

5.6.2. Tensile testpieces. The testpieces for tensile tests are generally circular (see Figs 5.10 and 5.11) or rectangular in cross-section, the diameters of the former usually being arranged to give a convenient cross-sectional area in order to make calculations easier. The gauge length $L_0 = 5\cdot65\sqrt{S_0}$ (5 diameters for round testpieces) where S_0 is the cross-sectional area along the gauge length.

(*Note:* For convenience of reference, the symbols used are those given in B.S. 18.)

Figs 5.12 and 5.13 show the differences in fracture between brittle and ductile materials.

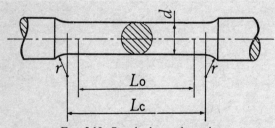

Fig. 5.10. Standard round testpiece

Fig. 5.11. Prepared testpiece

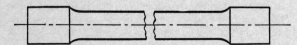

Fig. 5.12. Testpiece of brittle material after fracture

Fig. 5.13. Testpiece of ductile material after fracture

EXAMPLE 5.7 In a tensile test on a certain metal specimen the following results were obtained:

original gauge length $= 56\cdot5$ mm,
final gauge length $= 72\cdot5$ mm,
original cross-sectional area $= 100$ mm²,
cross-sectional area at fracture $= 50$ mm²,
load at yield point $= 21$ kN,
maximum load $= 48$ kN.

Estimate: (a) the yield stress,
(b) the tensile strength of the material,
(c) the percentage elongation, and
(d) the percentage reduction in area.

SOLUTION

(a) Yield stress $= \dfrac{\text{load at yield}}{\text{original cross-sectional area}}$

$= \dfrac{21\ 000\ \text{N}}{100\ \text{mm}^2}$

$= 210\ \text{N/mm}^2 = 210\ \text{MN/m}^2$ or MPa.

(b) Tensile strength $= \dfrac{\text{maximum load}}{\text{original cross-sectional area}}$

$= \dfrac{48\ 000\ \text{N}}{100\ \text{mm}^2}$

$= 480\ \text{N/mm}^2 = 480\ \text{MN/m}^2$ or MPa.

(c) Percentage elongation $= \left(\dfrac{l_f - l_0}{l_0}\right) \times 100$ per cent

$= \left(\dfrac{72\cdot5\ \text{mm} - 56\cdot5\ \text{mm}}{56\cdot5\ \text{mm}}\right) \times 100$ per cent

$= \dfrac{16}{56\cdot5} \times 100 = 28\cdot3$ per cent.

(d) Percentage reduction in area $= \left(\dfrac{A_0 - A_f}{A_0}\right) \times 100$ per cent

$= \left(\dfrac{100\ \text{mm}^2 - 50\ \text{mm}^2}{100\ \text{mm}^2}\right) \times 100$ per cent

$= \dfrac{50}{100} \times 100 = 50$ per cent.

5.7. PROOF STRESS

Not all materials have a yield point as is shown by the curves in Fig. 5.9. In such cases a yield stress cannot be found and some suitable substitute must be obtained. This is found by using the load–extension curve to estimate the load, known as the *proof load*, corresponding to a given non-proportional extension of the testpiece.

The stress produced by the proof load is known as the *proof stress* and is defined as the stress at which a small non-proportional extension of the testpiece is produced corresponding to a specified percentage of the gauge length. The specified amount of non-proportional extension varies according to the design requirements but is usually 0·1 per cent or 0·2 per cent.

The method of finding the proof load and hence the proof stress is shown in Fig. 5.14. Suppose that in a certain design a permanent strain of 0·1 per cent is to be allowed, then the corresponding 0·1 per cent proof stress would be determined. The required 0·1 per cent of the gauge length is first calculated and marked on the extension axis at point *P*. A line *PQ* is carefully drawn parallel to the linear portion of the load–extension curve and from point *Q* on the curve the proof load is read off the load axis at point *R*. Then

$$\text{proof stress} = \frac{\text{proof load}}{\text{original cross-sectional area}}.$$

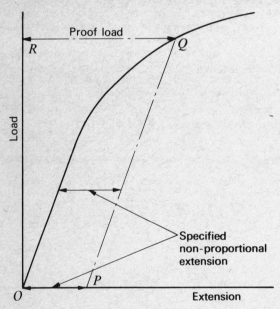

FIG. 5.14

SUMMARY

$$\text{Stress} = \frac{\text{normal force}}{\text{area}}.$$

The SI unit of stress is the pascal (Pa).

Typical derived units of stress are: N/m^2, N/mm^2, MN/m^2.

$$1\ N/mm^2 = 1\ MN/m^2\ \text{or MPa}.$$

$$\text{Strain} = \frac{\text{deformation}}{\text{original dimension}}.$$

Strain has no units: it is a ratio and is therefore expressed as a number.

$$\text{Spring stiffness} = \frac{\text{force}}{\text{deflection}}.$$

Stiffness is measured in units such as N/mm.

$$\text{Modulus of elasticity} = \frac{\text{direct stress}}{\text{direct strain}}.$$

$$\text{Modulus of rigidity} = \frac{\text{shear stress}}{\text{shear strain}}.$$

The units for the elastic constants may be the same as for stress, but the preferred unit is the kN/mm^2 or GPa.

$$\text{Tensile strength} = \frac{\text{maximum load}}{\text{original cross-sectional area}}.$$

$$\text{Factor of safety} = \frac{\text{tensile strength}}{\text{maximum allowable working stress}}.$$

Factor of safety is a ratio and has no units.

$$\text{Proof stress} = \frac{\text{proof load}}{\text{original area}}.$$

EXERCISES 5

1. A tie bar has a cross-sectional area of 320 mm² and carries a load of 40 kN. Under this load its length increases by 1 mm, over an initial length of 2 m.
   (a) The stress in the bar is...
   (b) The strain in the bar is...

Most relevant section: 5.3

U.E.I.

2. Hooke's law states that the strain in a material is directly proportional...within
   ....
   Stress can be measured in units of...............................

Most relevant sections: 5.3 and 5.4

U.E.I.

3. A 10 N load is hanging from a wire. If the stress in the wire is limited to 7 MN/m² or MPa the minimum area of cross-section is...

Most relevant section: 5.3

U.E.I.

4. A tie bar having a 25 mm square cross-section is 1·6 m long. When carrying a load of 40 kN its extension is 0·5 mm. Another tie bar of 20 mm square cross-section and of the same material has a length of 1 m, and carries a load of 25 kN. What will be the extension of this second tie bar?

Most relevant sections: 5.3 and 5.4.2

U.E.I.

5. State Hooke's law.
In an experiment on a 25 mm diameter bar 3 m long, a tensile load of 160 kN produces an extension of 9 mm. Calculate the stress and the strain in the bar. What would be the load when the extension is 2 mm?

Most relevant sections: 5.3 and 5.4.2

W.J.E.C.

6. Sketch a typical stress–strain graph obtained in a tensile test on mild steel. Show on it

   (i)   the limit of proportionality,
   (ii) the yield stress, and
   (iii) the tensile strength.

Most relevant section: 5.6.1

C.G.L.I.

7. Define (a) a tensile force and (b) a compressive force applied to a material. An extension of 0·4 mm occurred in a 3 m rod when subjected to a load. Calculate the strain produced in the rod.

Most relevant sections: 5.1 and 5.3

Y.C.F.E.

8. (a) A cylinder is to have a bore of 140 mm and the end cover plate will be secured by studs. If the pressure inside the cylinder is to be 1·8 MPa, calculate the force, in newtons, that will tend to separate the cover plate from the cylinder.
   (b) If the cover plate is to be secured by six studs and the stress in the plain portion of a stud is not to exceed 42 N/mm² or MPa, calculate the minimum cross-sectional area of a single stud.
   (c) If stud sizes are available with diameters in increments of whole numbers of millimetres, determine a suitable stud diameter.

Most relevant sections: 1.5 and 5.3

C.G.L.I.

9. State Hooke's law.
A helical spring extends 50 mm when a load of 80 N is applied. What would you expect the extension to be if an additional 10 N were applied?

Most relevant section: 5.4.1

E.M.E.U.

10. What will be the extension produced in a wire 2·5 m long in which the tensile stress is 100 MN/m² or MPa and for which the value of E is 200 kN/mm² or GPa?

Most relevant sections: 5.3 and 5.4.2

U.E.I.

11. In Fig. 5.15 ABC is a lever hinged at A and supported in a horizontal position by a vertical steel wire BD which is 1·5 m long, 1·2 mm diameter and rigidly fixed at D. $AC = 600$ mm, $AB = 200$ mm.
By how much will the wire stretch when a casting weighing 100 N is suspended from C and what downward movement will this cause at C?
E for steel is 210 kN/mm² or GPa.

Most relevant sections: 3.4, 5.3, and 5.4.2

C.G.L.I.

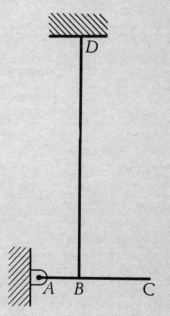

FIG. 5.15

12. Explain the meaning of the terms: (a) stress and (b) strain. A steel tie bar of rectangular cross-section, 50 mm × 5 mm is 2 m long. Calculate the load carried by the bar if the extension produced is 0·5 mm and if Young's modulus for the material is 210 kN/mm² or GPa. Determine also the tensile strength of the material if under the above conditions the factor of safety is 8.

Most relevant sections: 5.3, 5.4.2, and 5.5.1

Y.C.F.E.

13. Explain the term 'factor of safety'.
A tie bar is to withstand a maximum tensile load of 100 kN. The tensile strength is 600 MN/m² or MPa and the factor of safety is to be 6.

Calculate:

(a) the diameter of the bar;
(b) the extension on a length of 1·75 m when the load in the rod is 100 kN.
$$E = 200 \text{ kN/mm}^2 \text{ or GPa.}$$

Most relevant sections: 5.3, 5.4.2, and 5.5.1

E.M.E.U.

14. Describe in detail an experiment which may be carried out to determine the relationship between the load applied and the extension produced in a material to which a tensile load is applied.
In such an experiment on a wire 1·6 mm diameter, a load of 200 N produced an extension of 0·915 mm over a length of 1·9 m. Calculate the stress, strain, and the modulus of elasticity for the material in kN/mm² or GPa.

Most relevant sections: 5.3 and 5.4.2

E.M.E.U.

15. State Hooke's law.
A mild steel bar 250 mm long has a diameter of 20 mm for 150 mm of its length and is 30 mm diameter for the remainder. It is subjected to a direct pull of 20 kN. Given that Young's modulus (*E*) is 210 kN/mm² or GPa, determine:

(a) the stress in each part of the bar, and
(b) the total extension.

Most relevant sections: 5.3 and 5.4.2

U.E.I.

16. A wire of diameter 3 mm and length 1·5 m is loaded in tension with a load of 600 N and the extension is found to be 1·1 mm.
Determine for the material (a) the stress, (b) the strain, and (c) Young's modulus.

Most relevant sections: 5.3 and 5.4.2

U.L.C.I.

17. A steel bar has a cross-section 15 mm by 20 mm and its length is 250 mm. If Young's modulus for the steel is 210 kN/mm² or GPa and the tensile strength is 420 N/mm² or MPa, calculate the maximum load that can be carried using a factor of safety of 3. What will be the extension under this load?

Most relevant sections: 5.3, 5.4.2, and 5.5.1

W.J.E.C.

18. In a tensile test on a certain metal specimen the following results were obtained:

original diameter of testpiece	= 11·3 mm,
original gauge length	= 56·5 mm,
diameter of testpiece at fracture	= 8·0 mm,
final gauge length	= 72·5 mm,
yield load	= 21 kN,
maximum load	= 48 kN.

Determine:

(a) the yield stress,
(b) the tensile strength,
(c) the percentage elongation,
(d) the percentage reduction in area.

Most relevant section: 5.6.1

U.E.I.

19. Fig. 5.16 shows a shouldered pin subjected to a tensile force of 7700 N. The tensile stress in the shank of

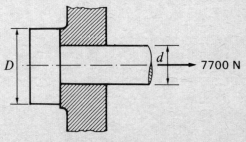

Fig. 5.16

the pin is not to exceed 50 N/mm² or MPa and the compressive stress on the underside of the head is not to exceed 20 N/mm² or MPa. Determine the dimensions *d* and *D*. (Take π as $\frac{22}{7}$.)

Most relevant section: 5.3

C.G.L.I.

20. (a) In connection with the strength of materials, explain why a proof stress is often required.
   (b) The following results were obtained from a tensile test on a steel specimen of diameter 10 mm and gauge length 50 mm:

   Load (kN)   5·25 10·50 15·75 21·00 26·25
      30·94 34·71 36·82 38·03 38·93 40·00
   Extension (mm) 0·02 0·04 0·06 0·08 0·10
      0·12 0·14 0·16 0·18 0·20 0·22

Plot the load–extension graph and determine the 0·1 per cent proof stress.

Most relevant section: 5.7

U.E.I.

ANSWERS TO EXERCISES 5

1. (a) 125 N/mm² or MPa, (b) 0·000 5. 3. 1·43 mm². 4. 0·305 mm. 5. 325 N/mm² or MPa, 0·003, 35·6 kN. 7. 0·000 13. 8. (a) 27·6 kN, (b) 110 mm², (c) 12 mm. 9. 56·25 mm. 10. 1·25 mm. 11. 2 mm, 6 mm. 12. 13·125 kN, 420 MN/m² or MPa. 13. (a) 35·6 mm, (b) 0·875 mm. 14. 99·55 N/mm² or MPa, 0·000 48, 207 kN/mm² or GPa. 15. (a) 63·5 N/mm² or MPa, 28·3 N/mm² or MPa, (b) 0·044 mm. 16. (a) 85 N/mm² or MPa, (b) 0·000 73, (c) 116 kN/mm² or GPa. 17. 42 kN, 0·171 mm. 18. (a) 210 MN/m² or MPa, (b) 480 MN/m² or MPa, (c) 28·3 per cent, (d) 49·7 per cent. 19. 14 mm, 26·2 mm. 20. 496·5 N/mm² or MPa.

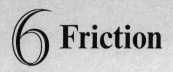 Friction

6.1. FRICTION

Whenever two surfaces slide or roll over one another the motion is opposed by a resistance known as *sliding (or kinetic) friction*. Even when there is no motion there is dormant resistance known as *static friction*. Static friction is generally greater than sliding friction: that is why more force is needed to set up motion than to maintain it at a uniform speed.

6.2. DRY FRICTION

The term *dry* or *solid* friction implies that there is no lubricant between the surfaces sliding or rolling over one another. In practice, perfectly dry friction is rare, because there is often a film of oxide or some other contaminant covering the surfaces and this gives some amount of lubrication.

Although the surface of a piece of material may appear to be perfectly smooth, it is in fact impossible to obtain a surface which is completely free from irregularities. When two materials are placed together they will be supported on the surface peaks (Fig. 6.1) so that the actual area of contact is very small. The intensity of pressure at the peaks is thus extremely high, and this causes plastic flow which results in adhesion. Friction is therefore due to the formation of junctions between the materials at the peaks, and it is these junctions which have to be sheared in order that motion can take place.

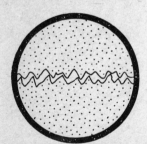 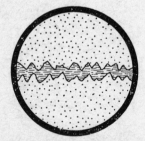

FIG. 6.1. Unlubricated surfaces   FIG. 6.2. Lubricated surfaces

6.2.1. Lubrication. Friction and wear can be greatly reduced if the surfaces of the materials are separated by a suitable lubricant (Fig. 6.2), generally an oil. If separation is complete, dry friction is eliminated, and there is only the *fluid* friction, or *viscosity*, of the oil which has to be overcome.

As well as reducing friction and wear, lubrication also serves to protect the materials against corrosion, and to dissipate excessive heat.

The most commonly employed lubricants are oils and greases, while molybdenum disulphide is used in many applications where conventional lubricants are inadequate.

6.2.2. Laws of dry friction. The amount of friction between two surfaces is governed by the following factors:

1. The roughness or smoothness of the surfaces.
   (Where it is necessary to reduce friction, the surfaces are made as smooth as possible. Conversely, if friction is desirable the surfaces may be roughened.)
2. The normal reaction between the two surfaces (i.e. the reaction at right angles between the surfaces).
3. The kind of materials in contact.
   (Surfaces of various materials may be prepared in the same manner, but the amount of friction present when the materials are placed together may be different, e.g., steel will slide more easily on nylon than it will on rubber.)

Note: Friction is independent of the area of the surfaces in contact.

6.3. ADVANTAGES AND DISADVANTAGES OF FRICTION

In engineering, friction has many advantages and disadvantages. By careful selection of materials and preparation of surfaces the amount of friction between the surfaces can be greatly increased or decreased. However, even when using the finest materials and lubricants, friction cannot be entirely eliminated and for this reason no machine can be said to be 'frictionless.'

Some examples of the many advantageous and disadvantageous effects of friction are given below.

1. Advantages:
   brakes and clutches,
   belt and pulley drives,
   clamping devices.
2. Disadvantages:
   shaft running in bearings,
   machine-tool slides,
   brushes on commutators and slip-rings.

6.4. COEFFICIENT OF FRICTION

The coefficient of friction (symbol μ) for a pair of materials in contact is given by

$$\mu = \frac{\text{friction force } R}{\text{normal reaction between surfaces } N}.$$

R and N are expressed in the same units, and as a result μ is a ratio, having no units of its own.

Consider the following cases where a block of weight W is pulled along a horizontal plane by a force of magnitude F. *Case 1* (see Fig. 6.3).

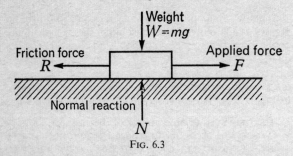

FIG. 6.3

While the block moves at a uniform speed, or while it is stationary, a resistance is experienced which is equal and opposite to F. The resistance is the friction force R, which always opposes motion.

Thus in this case,

$$\mu = \frac{R}{N} = \frac{F}{W}.$$

If the force F causes the block to accelerate, it is clear that F will be greater than the friction force R. The effect of acceleration will be considered in Chapter 8.

Case 2: In this case the force F is applied inclined at an angle θ to the horizontal plane. It is seen from Fig. 6.4 that the friction force R will be equal and opposite to the horizontal component $F \cos \theta$ of the applied force. Furthermore, the vertical component $F \sin \theta$ is opposing W and hence reduces the normal reaction N between the surfaces.

Thus, in this case,

$$\mu = \frac{R}{N} = \frac{F \cos \theta}{W - F \sin \theta}.$$

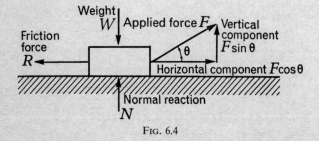

FIG. 6.4

Case 3: From the foregoing it is clear that if the applied force F is inclined down towards the horizontal plane, the normal reaction N will be increased.

Thus, in this case,

$$\mu = \frac{R}{N} = \frac{F \cos \theta}{W + F \sin \theta}.$$

TABLE 6.1
Average values of coefficient of friction

Materials (dry)	μ
Steel on cast iron	0·20
Steel on brass	0·15
Cast iron on brass	0·15
Leather on cast iron	0·55
Brake lining on cast iron	0·60
Rubber on asphalt	0·65
Rubber on concrete	0·70

6.5. ANGLE OF FRICTION

The forces R and N may be added vectorially to give the resultant reaction of the surface, since they are both exerted on the sliding block due to contact with the horizontal plane.

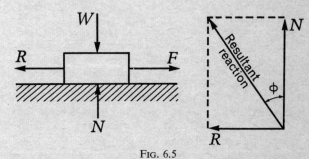

FIG. 6.5

From Fig. 6.5 it is seen that since R and N are at right-angles,

$$\tan \phi = \frac{R}{N},$$

$$\text{but } \mu = \frac{R}{N},$$

$$\therefore \tan \phi = \mu.$$

The angle ϕ reaches a maximum value when R is a maximum. This *maximum value of angle ϕ* is known as the *angle of friction.*

EXAMPLE 6.1. A horizontal force of 2·5 N moves a block of metal along a level surface at a uniform speed. If the coefficient of friction is 0·125, calculate the weight of the block.

SOLUTION

In this problem the applied force and the surfaces are horizontal as shown in Fig. 6.6.

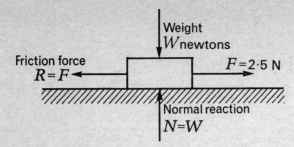

FIG. 6.6

Then $\mu = \dfrac{R}{N} = \dfrac{F}{W}.$

Thus weight $W = \dfrac{F}{\mu}$

$= \dfrac{2 \cdot 5}{0 \cdot 125}$

$= 20$ N.

EXAMPLE 6.2. A machine of mass 250 kg rests on a horizontal floor and is pulled into position by means of a rope inclined at 20° to the horizontal. If a tension in the rope of 500 N is just sufficient to move the machine, determine the coefficient of friction between the machine and floor. Take g as 9·81 m/s².

SOLUTION

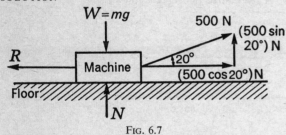

FIG. 6.7

In this problem the applied force is resolved into two components whose magnitudes are shown in Fig. 6.7. These components may also be found by employing a graphical method.

Weight of machine $W = mg = 250 \times 9 \cdot 81$ N

$= 2453$ N.

Coefficient of friction $\mu = \dfrac{R}{N} = \dfrac{500 \cos 20°}{2453 - 500 \sin 20°}$

$= \dfrac{500 \times 0 \cdot 9397}{2453 - 500 \times 0 \cdot 342}$

$= \dfrac{469 \cdot 9}{2453 - 171}$

$= \dfrac{469 \cdot 9}{2282}$

$= 0 \cdot 206.$

EXAMPLE 6.3. Fig. 6.8 shows a steel component weighing 100 N placed on the magnetic chuck of a surface grinding machine. Find the minimum magnetic force which must be provided by the chuck if the grinding operation produces a horizontal force of 48 N on the work. The coefficient of friction between component and chuck is 0·12.

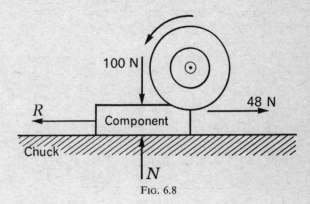

FIG. 6.8

SOLUTION

The friction force required will be equal to the horizontal force on the component.

∴ friction force $R = 48$ N.

From $\mu = \dfrac{R}{N},$

the normal reaction between chuck and work is

$N = \dfrac{R}{\mu} = \dfrac{48}{0 \cdot 12}$ N

$= 400$ N.

But the component weighs 100 N.

∴ magnetic force $= N - W$

$= 400 - 100$ N

$= 300$ N.

EXAMPLE 6.4. A steel slider was pulled at constant speed along a horizontal cast-iron surface during an experiment to determine the coefficient of friction for the materials. Weights W_1 were placed on the slider and the corresponding horizontal pulls P required to maintain motion were recorded as follows:

W_1 (N)	0·5	1·0	1·5	2·0	2·5	3·0
P (N)	0·35	0·46	0·54	0·65	0·74	0·85

Using these values, plot a graph and determine:

(a) the coefficient of friction, and
(b) the weight of the slider.

SOLUTION

The apparatus used during the experiment is shown in Fig. 6.9. The amount of friction at the pulley, being small, will be regarded as negligible.

The slope of the graph (Fig. 6.10) represents the coefficient of friction.

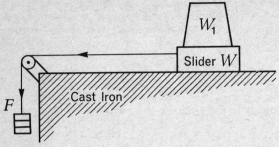

FIG. 6.9

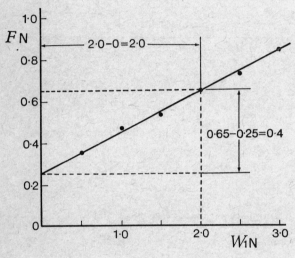

FIG. 6.10

From the graph:

$$\text{coefficient of friction } \mu = \frac{0.4}{2} = 0.2.$$

The graph is produced backwards to meet the vertical axis. When $W_1 = 0$, $F = 0.25$ N, i.e. the force required to move the slider only.

$$\text{Then from } \mu = \frac{R}{N} = \frac{F}{W},$$

$$\text{weight of slider } W = \frac{F}{\mu}$$

$$= \frac{0.25}{0.2} \text{ N}$$

$$= 1.25 \text{ N}.$$

SUMMARY

Frictional force always opposes motion.
Friction depends upon:

(1) condition of surfaces;
(2) normal reaction;
(3) nature of materials.

Friction is independent of contact area.

Coefficient of friction $\mu = \dfrac{\text{friction force } R}{\text{normal reaction } N}$.

Coefficient of friction has no units.

$\tan \phi = \dfrac{R}{N}$, where ϕ is the angle of friction.

Lubrication serves to:

(1) reduce friction and wear;
(2) protect the materials against corrosion;
(3) dissipate excessive heat.

EXERCISES 6

1. Define coefficient of friction.

Most relevant section: **6.4**

W.J.E.C.

2. If a lubricant is introduced between two sliding surfaces, what happens to the coefficient of friction?

Most relevant sections: **6.2.1** and **6.4**

U.L.C.I.

3. A horizontal force of 2·5 N will just move a body of weight 20 N along a level surface. Calculate the coefficient of friction.

Most relevant section: **6.4**

U.E.I.

4. Describe, with the aid of sketches, an experiment which can be performed to determine the coefficient of sliding friction between two flat surfaces. Explain how the effect of varying the load between the surfaces may be demonstrated, and state the conclusions which may be drawn from this experiment.

Most relevant section: **6.4**

U.E.I.

5. (a) A block weighing 8·5 N requires a horizontal force of 1·5 N to move it along a horizontal plane. Calculate the coefficient of sliding friction.

(b) By lubricating the sliding surfaces the coefficient of sliding friction is reduced by 80 per cent. What weight of block can now be moved by the same horizontal force?

Most relevant sections: **6.2.1** and **6.4**

U.E.I.

6. A horizontal force of 2·5 kN will just move a body of mass 1 t along a level surface. Calculate the coefficient of friction. Take g as 9·81 m/s².

Most relevant section: **6.4**

Y.C.F.E.

7. A 10 mm cube of steel is gripped in a vice until the force between the jaws is 300 N. Find

(i) the compression stress in the steel cube, and
(ii) the minimum force that will move the cube in the jaws against friction if $\mu = 0.3$.

Most relevant sections: **5.3** and **6.4**

C.G.L.I.

8. State two factors which affect the friction between two surfaces in rubbing contact. To move a scrap iron block of 300 kg which stands on a rough floor requires a horizontal force of 1 kN. Calculate the coefficient of friction. Take g as 9·81 m/s².

Most relevant sections: **6.2.2** and **6.4**

U.E.I.

9. A horizontal force of 1·8 kN is required to move a block of mass 500 kg along a level surface at a uniform speed. If the coefficient of friction is unchanged, determine the force necessary to move the block at the same uniform speed when the force is pushing the block and inclined at 45° into the plane. Take g as 9·81 m/s².

Most relevant sections: **6.2** and **6.4**

Y.C.F.E.

10. A machine, of mass 300 kg, is pulled along a horizontal floor by means of a rope inclined at 20° to the horizontal. If a tension in the rope of 600 N is just sufficient to move the machine, determine the coefficient of friction between the machine and the floor. Take the local gravitational acceleration g as 9·81 m/s².

Most relevant section: **6.4**

ANSWERS TO EXERCISES 6

3. 0·125. **5.** (a) 0·177; (b) 42·5 N. **6.** 0·255. **7.** (i) 3 N/mm² or MPa, (ii) 180 N. **8.** 0·339. **9.** 4 kN. **10.** 0·206.

7 Work, energy, and power

7.1. WORK

When a force is applied to a body causing it to move, the force is said to be doing *work*. Work is calculated as *the product of the distance moved by the body and the force on the body measured in the direction of motion*. The unit of work is formed from the units of distance and force, i.e. newton-metre (N m), and is called the joule (J).

Fig. 7.1 shows a block free to move along a horizontal surface. A horizontal force of magnitude F (newtons) moves the block a distance s (metres), and the work done is sF (joules).

Now consider Fig. 7.2. The block is acted upon by a force of magnitude F whose direction is inclined at angle θ to the horizontal. The component of F in the direction of motion is $F \cos \theta$, and the work done in moving the block through a distance s (metres) is $sF \cos \theta$ (joules).

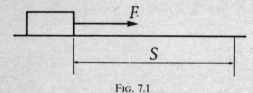

FIG. 7.1

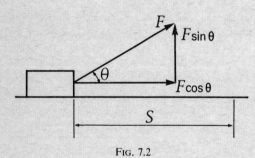

FIG. 7.2

EXAMPLE 7.1. A body rests on a horizontal surface. The pressure at the contact surface is 140 kPa and the contact surface area is 600 cm². The coefficient of friction at the contact surface is 0·25. What horizontal force is necessary to move the body, and how much work will be done in moving the body a distance of 70 m?

U.L.C.I.

SOLUTION

Total force on contact surface will be due to the weight of the body.

Weight W of body = pressure × surface area

$$= 140 \frac{\text{kN}}{\text{m}^2} \times \frac{600}{10^4} \text{ m}^2$$

$$\left(1 \text{ cm}^2 = \frac{1}{10^4} \text{ m}^2 \text{ and } 1 \text{ kPa} = 1 \text{ kN/m}^2\right)$$

$$= 8 \cdot 4 \text{ kN}.$$

Now coefficient of friction $\mu = \dfrac{R}{N}$

$$= \frac{F}{W} \text{ (in this case).}$$

Then horizontal force $F = \mu W$

$$= 0 \cdot 25 \times 8 \cdot 4 \text{ kN}$$

$$= 2 \cdot 1 \text{ kN}.$$

Work done = distance moved × horizontal force

$$= 70 \text{ m} \times 2 \cdot 1 \text{ kN}$$

$$= 147 \text{ kJ}.$$

7.1.1. Work diagrams. An amount of work may be represented by a diagram drawn to scale.

If a constant force F moves a body through a distance s, where F and s are in the same direction, the operation is depicted graphically by Fig. 7.3.

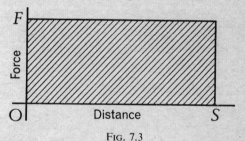

FIG. 7.3

The shaded section is a rectangle s units by F units and the area is sF square units, which represents the total work done.

The fact that the area of a force–distance diagram represents the work done is particularly useful when the force does not remain constant.

Where possible, the area of a work diagram is obtained by dividing the section into convenient shapes as shown by Fig. 7.4. For a diagram such as Fig. 7.5, the mid-ordinate rule (area = width of 1 strip × sum of mid-ordinates), or similar method, may be employed.

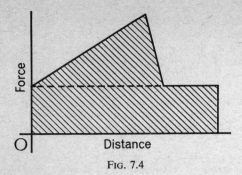

FIG. 7.4

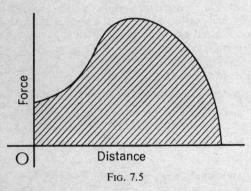

FIG. 7.5

EXAMPLE 7.2. A broach has a stroke of 0·8 m. The resistance offered by the workpiece is variable for the first 0·6 m of the stroke, rising uniformly from 1000 N to 2000 N. For the final 0·2 m of the stroke the resistance is constant at 2000 N. Calculate the work done during the stroke and the average resistance offered by the work-piece.

SOLUTION

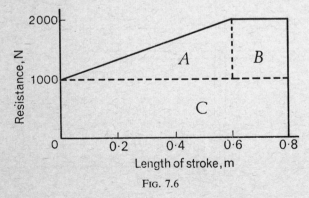

FIG. 7.6

Fig. 7.6 shows the work diagram which is divided into convenient sections *A*, *B*, and *C*.

The total work done during the stroke is numerically equal to the total area of the diagram. Then

total work done = area *A* + area *B* + area *C*
$$= (300 + 200 + 800) \text{ J}$$
$$= 1300 \text{ J}.$$

But, work done = distance moved × force,
(i.e. length of stroke) (i.e. resistance)

∴ average resistance = $\dfrac{\text{work done (J)}}{\text{length of stroke (m)}}$

$$= \frac{1300 \text{ J}}{0·8 \text{ m}}$$

$$= 1625 \text{ N}.$$

7.2. ENERGY

Anything capable of doing work is said to possess *energy*. Energy exists in various forms and may be considered as stored work awaiting use.

The following are some of the more common forms of energy.

1. Chemical energy—this form of energy is released or absorbed when a chemical reaction occurs. A chemical reaction is one in which various substances combine to produce other, distinctly different, substances. The energy is released by, for example, combustion of fuels or the electrolysis of a solution.

2. Atomic energy is liberated by splitting or combining parts of atoms. The splitting is called fission and is often produced from uranium in a nuclear reactor. The combining is called fusion. This is the process which makes the sun give out light and heat.

3. Heat energy exists by virtue of the motion of the molecules (smallest particles of substances which can have an independent physical existance) of which matter is made up. As the rate of motion increases a substance becomes hotter. Heat energy, in addition to raising the temperature of a substance, is able to change solids to liquids and liquids to gases.

4. Electrical energy can be produced in many ways, for example, through magnetism, from the windings of a generator or alternator, through thermoelectric effects, or electrochemically from a cell.

5. Mechanical energy—which is of three distinct types: (i) potential energy, (ii) kinetic energy, and (iii) strain energy.
   (i) Potential energy is possessed by a body by virtue of its position. For example, work is done in raising the hammer of a pile-driver, and when in its raised position the hammer stores this work as potential energy.
   (ii) Kinetic energy is possessed by a body by virtue of its velocity. For example, when the hammer of the pile-driver is allowed to fall, the potential energy is reduced because the height of the hammer above the end of the pile being driven gets less. The potential energy is being converted into kinetic energy since the hammer is now moving. At the instant of impact the potential energy is all converted into kinetic energy

which is now used to do work, i.e. to drive in the pile.
(iii) Strain energy may be stored in a piece of strained
elastic material. For example, when a bar is stretched
by a tensile force, work is done on the bar, which,
provided that the strain is kept within the elastic range,
is stored up in the bar as strain energy. This energy
can be regained on removal of the straining force.

7.2.1. Conservation of energy. *Energy can neither be created nor destroyed, but is convertible from one form into another.* (Energy and matter are also interconvertible.)

Fig. 7.7 illustrates the principle of conservation of
energy and shows how energy may be changed in form.
During the conversion process there may be a loss of
energy due to the inefficiency of the equipment used.
However, this 'lost' energy is not destroyed but is
converted into a form other than that desired. For
example, energy may be used in doing work to overcome
friction in a machine, and this work is converted into heat
at the rubbing surfaces.

In a conventional power station the chemical energy
possessed by coal is converted into heat energy in a boiler,
and this heat is used to produce high-pressure steam
which drives a turbine. The revolving rotor, which
possesses mechanical (kinetic) energy because of its
motion, is used to drive an alternator, and the rapid
rotation of the alternator rotor generates electricity in
the coils or windings. The electrical energy is fed along
the power lines and is converted back into other forms
of energy as required, e.g. for electro-plating, electric
heaters, and electric motors.

In a nuclear power station the atomic energy of
uranium is released by fission in a reactor, and the heat
generated is used to produce the high-pressure steam
needed to drive the turbo-alternator.

EXAMPLE 7.3. A body of weight 50 N is lifted at a uniform speed to a height of 20 m above the ground.
Calculate the work done and the potential energy stored
in the body at this height. If the body is allowed to fall
freely, find the kinetic energy after it has fallen 12 m.
What is the potential energy of the body now?

SOLUTION

Work done = distance (i.e. height) × force (i.e. weight)
$$= 20 \text{ m} \times 50 \text{ N}$$
$$= 1000 \text{ J.}$$

The work done is now stored in the body as potential
energy.

∴ potential energy = work done = 1000 J.

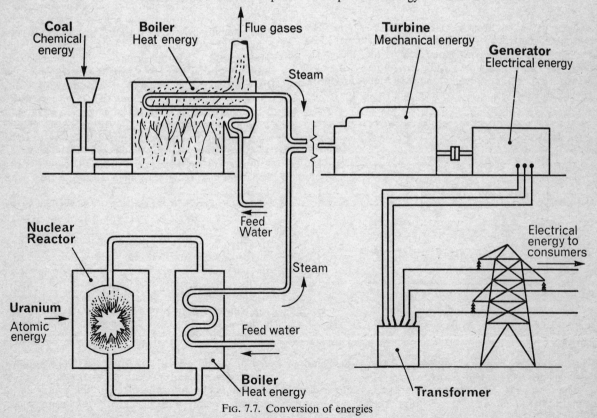

FIG. 7.7. Conversion of energies

After the body has fallen 12 m its height above the ground = 20 m − 12 m = 8 m.

∴ potential energy is now (8 m × 50 N) = 400 J.

As no external work is done during the fall, the total energy which the body possesses remains constant (see Fig. 7.8).

Potential energy + kinetic energy = constant.

Kinetic energy = original potential energy − present potential energy
= 1000 − 400 J
= 600 J.

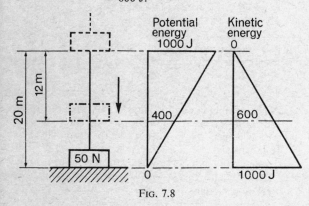

Fig. 7.8

7.3. POWER

Power is the rate at which work is done, i.e. the work per unit time.

$$\text{Power} = \frac{\text{work done}}{\text{time taken}} = \text{force} \times \text{velocity.}$$

Work can be done slowly or quickly depending upon the amount of power available. For example, two pumps may deliver equal amounts of water through the same distance, but one may require only half the time taken by the other. The pump working at the faster rate is therefore said to be twice as powerful as the other.

The unit of power is the watt (W).

From the above formula:

$$\text{Power (watts)} = \frac{\text{work (N m)}}{\text{time (s)}}$$

$$= \frac{\text{work (J)}}{\text{time (s)}} \quad \text{(since 1 N m = 1 J).}$$

Thus 1 watt = 1 joule per second
= 1 newton metre per second.

It will be seen that 1 kW = 1000 W
1 mW = 0·001 W, etc.

EXAMPLE 7.4. Water is pumped into a storage tank through a height of 15 m at a rate of 120 l/min. Calculate the power developed by the pump. Take *g* as 9·81 m/s².

E.M.E.U.

SOLUTION

Mass of 1 litre of water = 1 kg.
A litre of water weighs *mg* (newtons) = 1 × 9·81 = 9·81 N.
Rate of pumping = 120 l/min = 2 l/s,
∴ weight of water raised per second = 9·81 N × 2
= 19·62 N.

Work done per second = force × distance
(i.e. weight) (i.e. height)
= 19·62 N × 15 m
= 294·3 J.

Power developed by pump = work done per second,
= 294·3 W.

7.4. EFFICIENCY

Whenever motion occurs between two surfaces, a frictional force is set up which opposes the movement. There is also an opposition called 'windage' due to the air resistance on the moving parts. The work or power input to a machine must therefore be sufficient to overcome these resistances and to provide the necessary output from the machine.

A boiler used to generate steam requires the combustion of a fuel. Some of the heat liberated from the fuel will be absorbed by the boiler parts and surroundings so that not all of the heat is used in raising the steam.

The measure of the usefulness of a machine, or operation, is its efficiency.

$$\text{Efficiency} = \frac{\text{work (or power) output}}{\text{work (or power) input}}.$$

Efficiency can be expressed as a percentage, or as a fraction of unity (1) when it is called the per-unit efficiency.

EXAMPLE 7.5. The sand in a pit is raised 40 m to hoppers each holding 30 m³. It is raised by a conveyor system having a mechanical efficiency of 75 per cent by a motor developing 35 kW. How long will it take for a hopper to be filled if the relative density of the sand is 3·1?

The density of water is 1000 kg/m³.

Y.C.F.E.

SOLUTION

The relative density of the sand is 3·1 which means that a given volume of sand has a mass 3·1 times as much as the same volume of water.

Then mass of 1 m³ of the sand = 3·1 × 1000 kg
= 3100 kg, and
30 m³ of the sand has a mass of 93 000 kg.
Weight of sand = *mg* (where *g* = 9·81 m/s²)
= 93 000 × 9·81 N
= 912 330 N
= 912·33 kN.

Total work done to fill hopper
= force (i.e. weight) × distance (i.e. height)
= 912·33 kN × 40 m
= 36 493 kJ.

The motor develops 35 kW and 1 kW = 1 kJ/s,

∴ output of motor = input to conveyor = 35 kJ/s.

But the conveyor system has an efficiency of 75 per cent. Then work done per second by conveyor (i.e. output of conveyor)

$$= 35 \times \frac{75}{100} \text{ kJ}$$

$$= 26 \cdot 25 \text{ kJ}.$$

$$\text{Time to fill hopper} = \frac{\text{total work done}}{\text{work done/s}}$$

$$= \frac{36\ 493}{26 \cdot 25}$$

$$= 1390 \text{ s}$$

$$= 23 \text{ min } 10 \text{ s}.$$

SUMMARY

Work done = distance moved × force (in direction of motion).

Units of work are: Newton-metre (N m) or joule (J).

$$1 \text{ J} = 1 \text{ N m}.$$

An amount of work may be represented by the area under a force–distance graph drawn to scale.

Energy is the capacity for doing work.
Energy has the same units as work.
The principle of conservation of energy states that energy can neither be created nor destroyed, but is convertible from one form into another.

Power is the rate of doing work.

$$\text{Power} = \frac{\text{work done}}{\text{time taken}} = \text{force} \times \text{velocity}.$$

Typical units of power are: Watt (W), kW, MW, μW.

$$1 \text{ W} = 1 \text{ J/s}.$$

Efficiency is a measure of the usefulness of a machine or operation.

$$\text{Efficiency} = \frac{\text{work (or power) output}}{\text{work (or power) input}}.$$

Efficiency has no units, but is expressed as either a percentage or compared to unity.

EXERCISES 7

1. Write down two forms in which energy can appear. Give an example where each of these energy forms may be found.

Most relevant section: 7.2

U.L.C.I.

2. A force F moves through a distance L. The force is then increased to $2F$ which then moves through a further distance L. Sketch the work diagram.

Most relevant section: 7.1.1

U.L.C.I.

3. (a) What is the weight of an object if it requires a horizontal force of 250 N to move it along a horizontal surface when the coefficient of friction between the surfaces is 0·56?
(b) 4·46 kJ of work would be necessary to move the above object a distance of ....m.

Most relevant sections: 6.4 and 7.1

U.E.I.

4. A horizontal planing machine has a stroke of 1 m on the cutting stroke, the force on the cutting tool being 1·6 kN whilst on the return stroke the force is 0·2 kN. If the total time taken for these two operations is 5 s, determine the work done by the machine in 1 min.

Most relevant section: 7.1

U.L.C.I.

5. A lift carries 12 men from the ground floor to the fourth floor where 2 men get out. Six more men get out at the fifth floor and the remainder at the sixth floor. Calculate the work done by the lift during its upward journey, and draw a work diagram of force (scale: 1 mm = 100 N) on a base of distance (scale: 1 mm = 0·25 m) on graph paper. Take the mass of each man as 75 kg, the mass of the lift as 1000 kg, the height of each floor as 3 m and g as 9·81 m/s².

Most relevant section: 7.1.1

U.E.I.

6. A 25 kg block of steel is moved by the application of a horizontal force a distance of 0·5 m on a horizontal surface plate. How much less work would be done in moving the block over a lubricated surface plate, if the coefficient of friction for steel on dry cast iron is 0·2 and 0·03 when lubricated. Take g as 9·81 m/s².

Most relevant sections: 7.1 and 6.4

U.E.I.

7. (a) The diameter of a pump piston is 120 mm and the pressure in the cylinder is 0·55 MPa. What is the force on the piston?
If the piston moves through a distance of 200 mm whilst the pressure remains constant at 0·55 MPa how many joules of work are done?
(b) How much work is done by pumping 25 000 l of oil through a vertical distance of 10 m?
The oil has a relative density of 0·9: 1 litre of water has a mass of 1 kg: g is 9·81 m/s².

Most relevant sections: 1.3.2, 1.5.2 and 7.1

U.E.I.

8. A machine, of mass 300 kg, is pulled along a horizontal floor by means of a rope inclined at 20° to the horizontal. If a tension in the rope of 600 N is just sufficient to move the machine, determine the coefficient of friction between the machine and the floor. Take the local gravitational acceleration g as 9·81 m/s². How much work will be done against friction in moving the machine a distance of 10 m?

Most relevant sections: 6.4 and 7.1

U.E.I.

9. A compression spring is used in a buffer stop assembly. The spring has a free length of 500 mm and compresses at a uniform rate of 20 mm per 800 N additional force. When assembled the spring is initially compressed to a length of 480 mm and when in operation it compresses to a length of 260 mm.
(a) Draw the force–distance graph to show the compression of the spring when in use.
(b) From the graph determine the work done on the spring during the operation. Use scales 20 mm : 1000 N (force); 20 mm : 50 mm (distance).

Most relevant section: 7.1.1

U.E.I.

10. (a) State the SI unit for each of the following:
(i) force, (ii) work, (iii) energy, (iv) power.
(b) During a broaching operation a constant force of 18 kN was employed to draw the broach through the hole. The stroke of the machine is 1·0 m and the machine operates at 10 cutting strokes per minute. Determine:
 (i) the work done per cutting stroke,
 (ii) the power used by the machine,
 (iii) the efficiency, if the input power is 5·0 kW.

Most relevant sections: 7.1, 7.3, and 7.4

U.E.I.

11. Draw the load–extension graph of a spring which gives a total extension of 50 mm when the load applied increases from 0 N to 100 N. Calculate the work done in extending the spring.

Most relevant section: 7.1.1

Y.C.F.E.

12. A casting of mass $\frac{1}{2}$ t is raised at a uniform speed by a crane through 11 m vertically in $\frac{1}{4}$ min.
At what power is the crane working?
Take g as 9·81 m/s².

Most relevant sections: 7.1 and 7.3

C.G.L.I.

13. A compression spring of stiffness 4 N/mm has a free length of 80 mm. How much work is done in compressing it from 80 mm to 70 mm?

Most relevant sections: 5.4.1 and 7.1

C.G.L.I.

14. Define power and state the units in which it can be measured. A piston moves at a uniform velocity of 6 m/s against a resistance of 250 N. Find the power developed.

Most relevant sections: 7.1 and 7.3

E.M.E.U.

15. Convert 1 MJ into joules and show that 1 N m/s = 1 W. An electric haulage gear exerts a force of 3·5 kN over a distance of 12 m in 5 s. Find the work done per second in joules and the power input to the electric motor if its efficiency is 60 per cent.

Most relevant sections: 7.3 and 7.4

E.M.E.U.

16. Define work and power.
A lift cage of mass 1·5 t is raised steadily through a height of 10 m. The rope has a mass of 5 kg per metre length. Draw a diagram representing the work done and then calculate its numerical value. If the lift is accomplished in 9 s, what is the average power developed in lifting? Neglect friction losses and take g as 9·81 m/s².

Most relevant sections: 7.1.1 and 7.3

U.E.I.

17. A body moves under the influence of a variable force as follows:

Force (N)	23	26·5	32	35	37	37
Distance (m)	0	0·5	1·0	1·5	2·0	2·5
Force (N)	36·5	34·5	32·5	31·5	34	
Distance (m)	3·0	3·5	4·0	4·5	5·0	

Determine the average force and the work done in moving the body through 5 m.

Most relevant section: 7.1.1

U.L.C.I.

18. Water is raised by a pump through a height of 50 m, the efficiency of the pump being 55 per cent. If the quantity raised is 7 m³/h, calculate the output of the motor driving the pump. (Mass of 1 m³ of water = 1000 kg; $g = 9·81$ m/s².)

Most relevant sections: 7.3 and 7.4

U.L.C.I.

19. The following table gives the total thrust T exerted on a piston at distances S measured from the beginning to the end of its stroke of 90 mm.

T(N)	330	330	330	250	200	167	143	125	111	100
S(mm)	0	10	20	30	40	50	60	70	80	90

Draw a graph between T and S and use this to determine the total work done on the piston during the complete stroke.

Most relevant section: 7.1.1

U.E.I.

20. For a coal-burning power station, explain in correct sequence the various stages through which the energy from the coal is converted to electrical energy, and mention the main losses of energy which occur.

Most relevant section: 7.2.1

U.E.I.

ANSWERS TO EXERCISES 7

3. (a) 446 N, (b) 10 m. **4.** 21·6 kJ. **5.** 313·4 kJ. **6.** 20·85 J. **7.** (a) 6·23 kN, 1246 J. (b) 2·21 MJ. **8.** 0·206, 5·64 kJ. **9.** 1144 J. **10.** (b)(i) 18 kJ, (ii) 3 kW, (iii) 60 per cent. **11.** 2·5 J. **12.** 3·6 kW. **13.** 0·2 J. **14.** 1·5 kW. **15.** 8400 J, 14 kW. **16.** 149·6 kJ, 16·62 kW. **17.** 33·08 N, 165·3 J. **18.** 1·73 kW. **19.** 18·6 J.

8 Velocity and acceleration

8.1. DEFINITIONS OF TERMS

8.1.1. Vector quantity. A vector quantity is one involving both magnitude and direction. Examples of vector quantities are velocity, acceleration, and force (see also section 2.12).

8.1.2. Scalar quantity. A scalar quantity involves magnitude or size only. Examples of scalar quantities are speed, volume, mass, and time.

8.1.3. Speed. The rate of change of distance, or

$$\text{speed} = \frac{\text{distance}}{\text{time}}.$$

Units of speed are:

$$\frac{\text{kilometre}}{\text{hour}} \equiv \text{km/h},$$

$$\frac{\text{metre}}{\text{second}} \equiv \text{m/s, etc.}$$

8.1.4. Velocity. The rate of change of distance, in a particular direction.

Units employed for velocity are as those for speed: km/h, m/s, etc.

Confusion often arises in distinguishing between the terms velocity and speed since they both appear to refer to the same thing.

Speed has *magnitude* only.

Velocity has *magnitude* and *direction*.

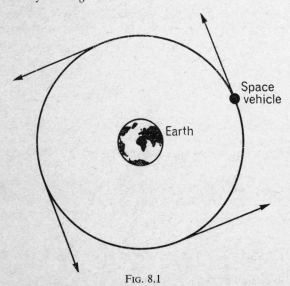

FIG. 8.1

Consider the case of a space vehicle orbiting the earth at 25 000 km/h (see Fig. 8.1).

The speed has a constant magnitude of 25 000 km/h. The velocity has a constant magnitude of 25 000 km/h but the direction of motion constantly changes. The velocity is therefore changing (but see section 9.6.2).

8.1.5. Uniform velocity. A body moving in a straight line and covering equal distances in equal times is said to have uniform velocity in a straight line.

8.1.6. Acceleration. The rate of change of velocity, or

$$\text{acceleration} = \frac{\text{change of velocity}}{\text{time taken}}.$$

Units of acceleration are $\frac{\text{m/s}}{\text{s}} \equiv \text{m/s}^2$, etc.

8.1.7. Uniform acceleration. When a body experiences equal changes of velocity in equal intervals of time it is said to have uniform acceleration.

8.1.8. Retardation or deceleration. The reverse process to acceleration. It is often referred to as negative acceleration.

8.2. UNIFORMLY ACCELERATED MOTION IN A STRAIGHT LINE

Let u be the initial or starting velocity (in m/s),
v the final velocity (in m/s),
a the acceleration (in m/s²),
s the distance travelled (in m),
and t the time taken (in s).

(a) Average velocity $= \dfrac{\text{initial velocity} + \text{final velocity}}{2}$

$$= \frac{u+v}{2}.$$

Distance travelled = average velocity × time,

$$\therefore s = \frac{(u+v)t}{2}. \tag{1}$$

(b) Consider a body travelling at an initial velocity of 44 m/s, and accelerating at 4 m/s² for 3 s. The velocity of the body after 1 s, 2 s, 3 s, and t s, will be as shown below.

After 1 s, velocity $= 44 + 1 \times 4 = 48$ m/s.
After 2 s, velocity $= 44 + 2 \times 4 = 52$ m/s.
After 3 s, velocity $= 44 + 3 \times 4 = 56$ m/s.
After t s, velocity $= 44 + t \times 4 = (44 + 4t)$ m/s.
$$\therefore v = 44 + 4t,$$

but 44 (m/s) = initial velocity u; and 4 (m/s²) = acceleration a,

$$\therefore v = u + at. \tag{2}$$

(c) From eqn (1), where $s = \dfrac{(u+v)t}{2}$

and replacing v by $u+at$ from eqn (2),

$$s = \frac{(u+u+at)t}{2}.$$

$$\therefore s = \frac{2ut+at^2}{2}.$$

and $s = ut + \frac{1}{2}at^2.$ \hfill (3)

(d) Transposing eqn (2), where $v = u+at$,

then $t = \dfrac{v-u}{a}.$

Using eqn (1) which is $s = \dfrac{(u+v)t}{2}$ and

substituting $t = \dfrac{v-u}{a}$,

$$s = \frac{u+v}{2} \times \frac{v-u}{a},$$

$$= \frac{uv+v^2-u^2-uv}{2a},$$

$$\therefore 2as = v^2-u^2,$$

or $v^2 = u^2 + 2as.$ \hfill (4)

The following points should be carefully noted when using the above equations:

(a) Each of these four equations connects four of the five quantities involved in problems on motion.
(b) Where a body starts from rest the initial velocity $u = 0$.
(c) When a body comes to rest the final velocity $v = 0$.
(d) Where a body is slowing down the retardation is $-a$.
(e) Constant systems of units must be used. For example if the velocities are expressed in m/s, then the distance must be expressed in m, time in s, and acceleration in m/s².

EXAMPLE 8.1
(a) Convert a speed of 72 km/h to m/s.

$$72 \text{ km/h} = 72 \times 1000 \text{ m/h}$$

$$= \frac{72 \times 1000}{60 \times 60} \text{ m/s},$$

$$\therefore 72 \text{ km/h} = 20 \text{ m/s}.$$

(b) Convert an acceleration of 50 km/min² into m/s².

$$50 \text{ km/min}^2 = 50 \times 1000 \text{ m/min}^2$$

$$= \frac{50 \times 1000}{60} \text{ m/s per min}$$

$$= \frac{50 \times 1000}{60 \times 60} \text{ m/s}^2$$

$$= 13 \cdot 9 \text{ m/s}^2.$$

An acceleration of 50 km/min² = 13·9 m/s².

EXAMPLE 8.2. A vehicle travels 100 m in 10 s while being accelerated at a constant rate of 0·5 m/s².

Calculate (a) its initial velocity,
(b) its final velocity, and
(c) the distance travelled during the first 5 s.

SOLUTION
(a) Quantities given are $s = 100$ m, $a = 0.5$ m/s², and $t = 10$ s.
It is required to find u.
The equation containing s, a, u, and t is

$$s = ut + \tfrac{1}{2}at^2.$$

Substituting, $100 = 10u + \frac{1}{2} \times \frac{1}{2} \times 10^2$,

so that $100 = 10u + \dfrac{100}{4}.$

$$100 - 25 = 10u,$$

$$u = \frac{75}{10} = 7\cdot5.$$

Initial velocity = 7·5 m/s.

(b) Final velocity.

Using $v = u + at$,
$$v = 7 \cdot 5 + 0 \cdot 5 \times 10,$$
$$\therefore v = 12 \cdot 5.$$

Final velocity = 12·5 m/s.

(c) Distance travelled in 5 s.

Using $s = ut + \frac{1}{2}at^2$,
$$s = 7 \cdot 5 \times 5 + \tfrac{1}{2} \times 0 \cdot 5 \times 5^2,$$
$$s = 37 \cdot 5 + 6 \cdot 25,$$
$$\therefore s = 43 \cdot 75.$$

Distance travelled in first 5 s = 43·75 m.

EXAMPLE 8.3. A shaping machine ram starts from rest and moves through a distance of 270 mm in $\frac{3}{4}$ s with uniform acceleration.

Calculate (a) the acceleration, and
(b) the velocity of the ram after moving through 270 mm.

SOLUTION
(a) In this example the ram starts from rest and therefore the initial velocity $u = 0$.
Known quantities are $s = 270$ mm, $t = \frac{3}{4}$ s, and $u = 0$.
Acceleration a is required.

Using $s = ut + \frac{1}{2}at^2$,
$$270 = 0 \times \tfrac{3}{4} + \tfrac{1}{2} a(\tfrac{3}{4})^2,$$
$$\therefore 270 = 0 + \frac{9}{32} a,$$

and $a = 270 \times \dfrac{32}{9} = 960.$

Acceleration = 960 mm/s².

(b) Final velocity.

$$\text{Using } v = u+at,$$
$$v = 0+960 \times \tfrac{3}{4},$$
$$\therefore v = 720.$$

Velocity after 270 mm = 720 mm/s.

EXAMPLE 8.4. A vehicle accelerates from rest with a uniform acceleration of 0.5 m/s^2 up to a velocity of 36 km/h. It travels at constant velocity for 1 km, the brakes are then applied and it comes to rest in 20 s. Calculate the total distance travelled and the total time taken.

U.L.C.I.

SOLUTION

In this question mixed units are given, i.e. km/h and m/s^2. Here the solution will be worked throughout in metres and seconds and therefore it is necessary for the velocity of 36 km/h to be expressed in m/s.

$$36 \text{ km/h} = \frac{36 \times 1000}{60 \times 60} \text{ m/s} = 10 \text{ m/s}.$$

$$\therefore 36 \text{ km/h} = 10 \text{ m/s}.$$

(a) To find distance travelled.
The motion is divided into three parts: (i) uniformly accelerated motion; (ii) motion at constant velocity; (iii) uniformly retarded motion.
(i) The known quantities are $u = 0$, $a = 0.5 \text{ m/s}^2$, and $v = 10 \text{ m/s}$.

It is required to find s.

Using $v^2 = u^2+2as$,

$$(10)^2 = 0+2 \times 0.5 \, s,$$
$$\therefore s = (10)^2, \text{ and}$$

distance travelled during uniform acceleration,

$$s = 100 \text{ (m)}.$$

(ii) When travelling at constant velocity, distance travelled = 1 km = 1000 m.
(iii) Known quantities are $u = 10 \text{ m/s}$, $t = 20 \text{ s}$, and $v = 0$. The unknown quantity is s.

Using $s = \left(\dfrac{u+v}{2}\right) t,$

$$\therefore s = \left(\frac{10+0}{2}\right) 20 = 100 \text{ m}$$

Distance travelled while retarding = 100 m,
$\therefore$ total distance travelled by vehicle

$$= 100 \text{ m}+1000 \text{ m}+100 \text{ m}$$
$$= 1200 \text{ m or } 1.2 \text{ km}.$$

(b) To find total time taken.
Again the motion is divided into three parts as before.
(i) Known quantities are $u = 0$, $a = 0.5 \text{ m/s}^2$, and $v = 10 \text{ m/s}$.

It is required to find t.

$$v = u+at,$$
$$10 = 0+0.5t.$$

$\therefore$ time taken to accelerate the vehicle,

$$t = \frac{10}{0.5} = 20 \text{ s}.$$

(ii) Time taken to travel 1000 m at 10 m/s is

$$t = \frac{1000}{10} = 100 \text{ s}.$$

(iii) Time given as 20 s for retarding the vehicle.
$\therefore$ total time taken for journey $= 20+100+20$ s
$$= 140 \text{ s}.$$

EXAMPLE 8.5. A sliding piece in a mechanism is brought to rest from a speed of 8 m/s in a distance of 1.5 m under uniform retardation.
Calculate (a) the retardation, and (b) the time taken.

SOLUTION

(a) Retardation. (*Note:* This example involves retardation and since the sliding piece is brought to rest, $v = 0$.)

Quantities given are $u = 8 \text{ m/s}$, $v = 0$, and $s = 1.5 \text{ m}$. It is required to find a.

Using $v^2 = u^2+2 \, as$

$$0 = 8^2+2a \times 1.5,$$
$$0 = 64+3a,$$
$$3a = -64,$$
$$\therefore a = -\frac{64}{3} = -21.33 \text{ m/s}^2.$$

Retardation = 21.33 m/s².

(b) Time taken.
Known quantities are
$v = 0$, $u = 8 \text{ m/s}$, and $a = -21.33 \text{ m/s}^2$.
The unknown quantity is t.

$$\text{Using } v = u+at,$$
$$0 = 8-21.33t,$$
$$-8 = -21.33t,$$

$$\therefore t = \frac{8}{21.33} = 0.375 \text{ s}.$$

Time required = 0.375 s.

8.3. MOTION UNDER GRAVITY

When a body falls freely under gravity it experiences an acceleration directed towards the centre of the earth.
This acceleration varies at different places on the earth's surface but for purposes of calculation its value is generally taken as 9.81 m/s^2.
The symbol g is used to denote this acceleration.
Since g is, within practical limits, a uniform acceleration the four equations already derived may be used to solve problems involving free falling bodies. In problems involving motion under gravity g replaces a in the equations of motion.

The following points should be noted when dealing with problems on motion under gravity.

(1) If a body is projected vertically upwards, at its greatest height, $v = 0$.
(2) When a body is projected upwards, it experiences a retardation equal in magnitude to g.
(3) The time for a body to reach its highest point and return to earth equals twice the time for the body to reach the greatest height.

EXAMPLE 8.6. A drop stamp falls freely from a height of 6 m under the action of gravity. Find its velocity at the moment of striking the tup ($g = 9·81$ m/s².)

SOLUTION

Drop stamp has initial velocity (u) = 0, height (s) = 6 m, and acceleration (a) = g = 9·81 m/s².
The final velocity is required.

$$\text{Using } v^2 = u^2 + 2gs,$$
$$v^2 = 0 + 2 \times 9·81 \times 6,$$
$$v^2 = 117·72$$
$$v = \sqrt{117·72}$$
$$\therefore v = 10·85.$$

The final velocity = 10·85 m/s.

EXAMPLE 8.7. With what vertical velocity must a jet of water emerge from a nozzle, if the jet is to reach a height of 40 m ($g = 9·81$ m/s².)

SOLUTION

Known quantities are: $s = 40$ m, $v = 0$, and $a = -g$ = $-9·81$ m/s².
It is required to find u.

$$\text{Using } v^2 = u^2 + 2gs,$$
$$0 = u^2 - 2 \times 9·81 \times 40,$$
$$-u^2 = -2 \times 9·81 \times 40,$$
$$u^2 = 784·8,$$
$$\therefore u = 28.$$

Vertical velocity = 28 m/s.

8.4. GRAPHICAL REPRESENTATION OF MOTION

Graphical solutions are particularly useful in cases where details of motion are obtained from experimental data and also in cases when the motion is made up of a number of distinct parts.
A number of typical graphs are now considered.

8.4.1. Distance–time graphs. On a distance—time graph the *velocity* at any instant is equal to the *slope* of the graph.
A straight-line graph (Fig. 8.2) has a constant slope. Hence the velocity of this motion is constant.

$$\text{Slope} = \frac{s}{t} = v.$$

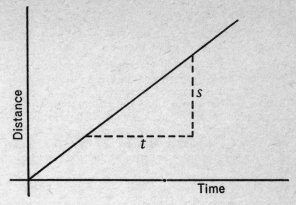

FIG. 8.2

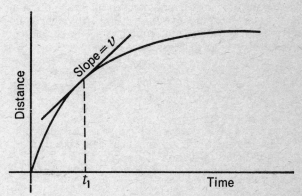

FIG. 8.3

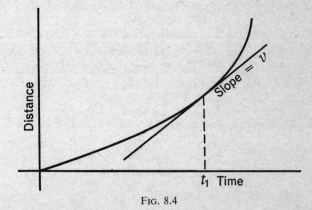

FIG. 8.4

Fig. 8.3 shows that as time increases velocity decreases. That is retarded motion. The horizontal portion of the graph indicates that as time increases the distance from the start remains constant, i.e. the body is at rest. Slope of the graph at t is a measure of the velocity at that time. Fig. 8.4 shows that with increasing time, the distance travelled increases more rapidly. This illustrates accelerated motion.

8.4.2. Velocity–time graphs

(1) The slope of a velocity–time graph measures acceleration.
(2) The area under a velocity–time graph determines distance travelled.
(3) Average velocity may be obtained by dividing the area under a graph by the corresponding time interval.

A graph of constant velocity (Fig. 8.5) shows:
Slope zero—no acceleration.

Distance travelled = vt.

Fig. 8.6 shows uniformly accelerated motion.

Slope constant—uniform acceleration.

Distance travelled = area of triangle = $\frac{1}{2}vt$.

The graph of uniformly retarded motion (Fig. 8.7) shows:

Slope constant and negative—uniform retardation.

Distance travelled = area of triangle = $\frac{1}{2}vt$.

Fig. 8.8 represents uniformly accelerated motion having initial velocity u, final velocity v, and acceleration a.

Hence distance travelled = area of triangle + area of rectangle or
$$s = \tfrac{1}{2}at^2 + ut.$$

Fig. 8·9 represents a motion experiencing variable acceleration.

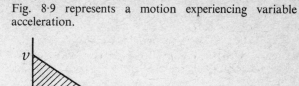

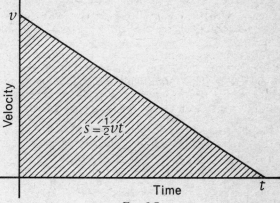

FIG. 8.7

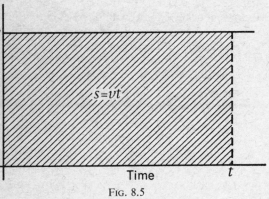

FIG. 8.5

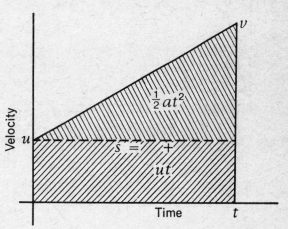

FIG. 8.8

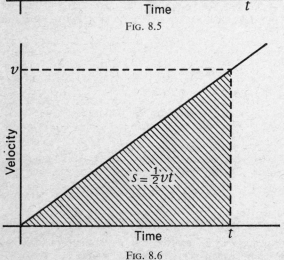

FIG. 8.6

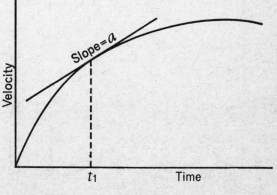

FIG. 8.9

Slope of graph at t represents the acceleration at time t. To determine the distance travelled in time t the area under the curve may be obtained by using the mid-ordinate rule.

EXAMPLE 8.8. A suburban train leaves station A and attains a speed of 8 m/s after 15 s: it is then further accelerated during the next 15 s until a speed of 20 m/s is reached.

The train maintains this speed, until just before entering station B, when the brakes are applied and the train is brought to rest in 33 s with uniform retardation. If the total running time from A to B is 183 s, sketch the velocity—time and distance—time graphs and determine:

(a) the distance between stations A and B, and
(b) the average speed of the train over this distance.

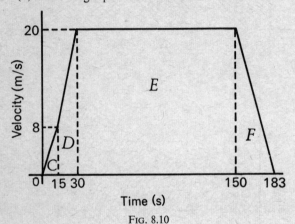

FIG. 8.10

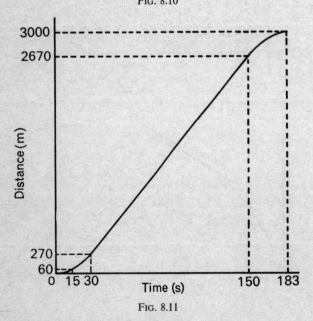

FIG. 8.11

SOLUTION

(a) Distance between A and B = area under graph shown in Fig. 8.10.

Area of triangle $C = \frac{1}{2} \times 15 \times 8 = 60$.

Area of trapezium $D = \frac{1}{2}(20+8) \times 15 = 210$.

Area of rectangle $E = 20 \times 120 = 2400$.

Area of triangle $F = \frac{1}{2} \times 33 \times 20 = 330$.

∴ total distance travelled = 3000 m.

Distance between stations A and B = 3 km.

$$\text{Average speed} = \frac{\text{total distance}}{\text{total time}} = \frac{3000 \text{ m}}{183 \text{ s}} = 16 \cdot 39 \text{ m/s}.$$

Consider the distance—time curve shown in Fig. 8.11.

Between 0 s and 15 s the curve is of parabolic form—uniformly accelerated motion (see Fig. 8.4).

Between 15 s and 30 s the curve is of parabolic form—uniformly accelerated motion.

Between 30 s and 150 s the path is a straight-line graph—constant velocity (see Fig. 8.2).

Between 150 s and 183 s the curve is of parabolic form—uniformly retarded motion (see Fig. 8.3).

EXAMPLE 8.9. The velocity v of a car at certain times is given in the table. Draw the velocity–time graph and find the distance travelled from $t = 3$ s to $t = 6$ s.

Time (s)	0	1	2	3	4	5	6
Velocity (m/s)	40	39	36	31	24	15	4

SOLUTION

The velocity–time graph is drawn to scale in Fig. 8.12, and the distance travelled is found by determining the area under the curve between the limits of $t = 3$ s and $t = 6$ s.

Using the mid-ordinate rule:

Area = width of 1 strip × sum of mid-ordinates.

Width of 1 strip = 1 (s).

Lengths of mid-ordinates are equal to 27·7, 19·7, and 9·7.

∴ area = 1 (27·7 + 19·7 + 9·7)

= 57·1.

Hence distance travelled = 57·1 m.

8.5. RESOLUTION OF VELOCITIES

Velocities may be resolved into two component velocities in a similar manner to forces (see Chapter 2).

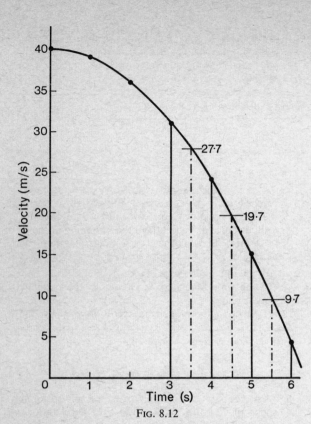

FIG. 8.12

The velocity *v* may be replaced by two component velocities at right-angles to each other as shown in Fig. 8.13.

Horizontal component of velocity $= v \cos \theta$.

Vertical component of velocity $= v \cos (90° - \theta)$
$= v \sin \theta$.

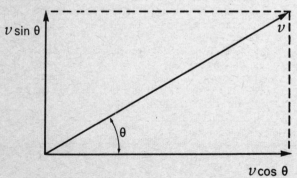

FIG. 8.13. Resolution of velocities

EXAMPLE 8.10. Calculate the rate of climb and forward velocity of an aircraft travelling at 400 km/h at an angle of 30° to the horizontal.

SOLUTION

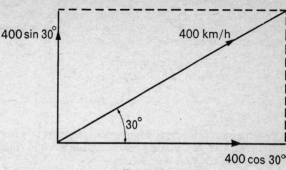

FIG. 8.14

See Fig. 8.14.

Rate of climb $=$ vertical velocity
$= 400 \sin 30°$ km/h
$= 400 \times 0\cdot5 = 200$ km/h.

Forward velocity $=$ horizontal velocity
$= 400 \cos 30°$ km/h
$= 400 \times 0\cdot866 = 346\cdot4$ km/h.

Rate of climb $= 200$ km/h

Forward velocity $= 346\cdot4$ km/h.

SUMMARY

Speed is the rate of change of distance and is a scalar quantity, i.e. it has magnitude only.

Velocity is the rate of change of distance and is a vector quantity, i.e. it has magnitude *and* direction.

Units of speed and velocity are km/h, m/s, etc.

Uniform velocity requires equal distances to be covered in equal intervals of time.

Acceleration is the rate of change of velocity.

Uniform acceleration indicates equal changes of velocity in equal intervals of time.

Units of acceleration are m/s², km/s², etc.

Retardation is the reverse of acceleration.

Equations of uniformly accelerated motion:

$$s = \left(\frac{u+v}{2} \right) t,$$
$$v = u + at,$$
$$s = ut + \tfrac{1}{2} at^2,$$
$$v^2 = u^2 + 2as,$$

where $v =$ final velocity, $u =$ initial velocity, $s =$ distance, $a =$ acceleration, and $t =$ time.

For motion under gravity: replace a by $-g$ for upward motion and a by $+g$ for downward motion in the above equations.

$$g = 9\cdot81 \text{ m/s}^2.$$

Distance–time graph: slope of *s–t* graph is a measure of velocity.

Velocity–time graph: area under v–t graph is a measure of distance travelled.

Slope of v–t graph gives acceleration at any instant. A velocity v may be resolved into two components at right-angles to each other:

$$\text{horizontal component} = v \cos \theta,$$
$$\text{vertical component} = v \sin \theta.$$

EXERCISES 8

1. An object falls freely from a platform which is at a height of 8 m. Calculate the time taken for it to reach the ground. Take $g = 9.81$ m/s^2.

Most relevant section: 8.3

N.C.T.E.C.

2. If a stone is allowed to fall freely from rest, how far will it drop in the first 2 s? Take $g = 9.81$ m/s^2 and neglect air resistance.

Most relevant section: 8.3

C.G.L.I.

3. A car travelling at 50 km/h accelerates uniformly to 70 km/h in 45 s. Sketch the velocity–time graph and calculate the distance travelled during this acceleration period.

Most relevant section: 8.4.2

Y.C.F.E.

4. A train accelerates uniformly from rest to 50 km/h in 5 min. Draw the velocity–time graph on the graph paper provided, and from this graph determine the acceleration in m/s^2.

Most relevant section: 8.4.2

E.M.E.U.

5. An aircraft is travelling 30° E of N at 700 km/h. What is its velocity resolved (a) due N; (b) due E?

Most relevant section: 8.5

U.E.I.

6. A car starts from rest and uniformly accelerates to a speed of 54 km/h in 20 s. Its speed then falls uniformly to 36 km/h. The brakes are then applied to bring the

car to rest in 3 s. The total distance travelled is 2000 m. Find the distance it moves when accelerating and when braking and the time taken for the whole journey.

Most relevant section: 8.2

Y.C.F.E.

7. A test run carried out on a locomotive and train when running along a level track showed that, starting from rest and accelerating uniformly, the locomotive reached its maximum speed after 5 km. Thereafter the speed remained constant until it had covered a total distance of 10 km from the beginning of the run in 9 min. Sketch the speed–time diagram, and determine:

(a) the maximum speed in km/h;
(b) the time taken to accelerate to the maximum speed; and
(c) the magnitude of the acceleration in m/s^2.

Most relevant sections: 8.2 and 8.4.2

Y.C.F.E.

8. A car accelerates uniformly from rest to a maximum speed of 72 km/h in 12 s. Find the acceleration in m/s^2 and the distance travelled in m.
Sketch the form of the velocity–time diagram which represents this accelerating period.

Most relevant sections: 8.2 and 8.4.2

U.E.I.

9. A train starts from rest and accelerates uniformly for 60 s, travels at uniform speed for the next 150 s and then retards uniformly to rest in 40 s, having travelled 4000 m. Find:
(a) the uniform speed; (b) the acceleration; (c) the retardation.
What time will be saved on the journey if the acceleration is increased by 100 per cent, the uniform speed and retardation remaining the same?

Most relevant section: 8.2

E.M.E.U.

10. A train has a speed of 40 km/h and 15 s later the speed has fallen to 30 km/h. Assuming uniform retardation
(a) draw a velocity–time graph to represent the motion.

(b) determine

    (i) the retardation,
    (ii) the distance travelled during that time.

Most relevant sections: 8.4.2 and **8.2**

N.C.T.E.C.

11. A vehicle starts from rest with constant acceleration of $1 \cdot 5 \text{ m/s}^2$ and this acceleration is maintained for 25 s. The velocity is maintained at a constant value for a period after which the vehicle is then brought to rest with uniform deceleration in 15 s. The total distance travelled is 1 km.

Draw the velocity–time diagram and use it to obtain:

(a) the constant velocity;
(b) the retardation;
(c) the total time for the journey.

Most relevant sections: 8.4.2 and **8.2**

U.L.C.I.

12. A vehicle of mass 800 kg travels in a straight line with a velocity v (m/s) after a time t (s) as given in the following table.

v(m/s)	0	2·5	5	7·5	7·5	7·5	6	4·5	3	1·5	0
t(s)	0	2	4	6	8	10	12	14	16	18	20

(a) Draw the velocity–time graph and use it to find
    (i) the acceleration of the vehicle,
    (ii) the deceleration of the vehicle, and
    (iii) the total distance travelled by the vehicle.

(b) Calculate the magnitude of the accelerating force.

Most relevant sections: 8.4.2 and **9.2**

U.E.I.

ANSWERS TO EXERCISES 8

1. 1·28 s. **2.** 19·62 m. **3.** 750 m. **4.** 0·046 m/s². **5. (a)** 606·2 km/h, **(b)** 350 km/h. **6.** 150 m, 15 m, 169·8 s. **7. (a)** 100 km/h, **(b)** 6 min, 0·077 m/s². **8.** 1·67 m/s², 120 m. **9. (a)** 20 m/s, **(b)** 0·33 m/s², **(c)** −0·5 m/s², 30 s. **10. (b) (i)** 0·185 m/s², **(ii)** 145·9 m. **11. (a)** 37·5 m/s, **(b)** 2·5 m/s², **(c)** 46·67 s. **12. (a)** 1·25 m/s², 0·75 m/s², 90 m, **(b)** 1000 N.

9 Force, motion, and torque

9.1. FORCE AND MOTION

The three fundamental laws of motion, as formulated by *Sir Isaac Newton* in the late seventeenth century, governing the behaviour of bodies under the action of forces are as follows:

1. *A body continues in a state of rest, or uniform motion in a straight line, unless acted upon by an external force.*
2. *The rate of change of momentum of a body is proportional to the applied force and takes place in the direction of the force*

This law may also be stated as follows:

If the resultant force acting on a body is not zero, the acceleration produced is proportional to the resultant force and takes place in the direction of that force.

From this law the following important relationship will be derived.

$$\text{Force} = \text{mass} \times \text{acceleration}.$$

3. *When bodies are in contact, action and reaction are equal and opposite.*

9.2. FORCE AND ACCELERATION

Momentum is defined as the product of mass and velocity, or

$$\text{momentum} = mv.$$

By Newton's second law: force is proportional to the rate of change of momentum (mv), so that if

F is the force, m the mass, and v the velocity,

$$F \propto \text{rate of change in } mv.$$

But m cannot change,

therefore $F \propto m \times$ the rate of change of v,

where the rate of change of velocity is acceleration a.

Hence

$$F \propto ma.$$

To change from a statement of proportionality to an equation requires the introduction of a constant k.

Thus

$$F = kma.$$

The simplest equation would be obtained if k were equal to 1.

This may be arranged by *choosing* a *unit* of force which would produce in *unit* mass a *unit* acceleration.

Substituting $F = 1$, $m = 1$ and $a = 1$ in the equation

$$F = kma$$
$$1 = k \times 1 \times 1$$

whence, $k = 1$.

The equation may now be stated as

$$F = m \times a.$$
$$\textbf{Force} = \textbf{mass} \times \textbf{acceleration}.$$

Note: In the use of this equation great care must be taken to use an absolute system of units. This means that the three units of length, mass, and time used are independent of the location on earth where measurements are made, and in fact would have the same significance if used on another planet.

In SI units:

unit force = 1 newton (N),
unit mass = 1 kilogram (kg),
unit acceleration = 1 metre per second per second (m/s²).

A newton is that force which, when acting on a mass of 1 kg produces an acceleration of 1 m/s².

Force (N) = mass (kg) × acceleration (m/s²).

EXAMPLE 9.1. Calculate the force required to produce an acceleration of 3 m/s² in a mass of 120 kg.

SOLUTION

$$\text{Using } F = ma$$
$$F = 120 \times 3$$
$$= 360.$$
$$\text{Accelerating force} = 360 \text{ N}.$$

EXAMPLE 9.2. Calculate the acceleration produced by (i) a force of 10 N acting on a mass of 12 kg, and (ii) a force of 20 kN acting on a mass of 50 t.

SOLUTION

(i) If the force is in newtons (N), and the mass in kilograms (kg), then the acceleration is in metres per second per second (m/s²).

$$\text{From } F = ma,$$
$$10 = 12 \times a,$$
$$\therefore a = \frac{10}{12} = \frac{5}{6}.$$

Acceleration $\qquad = \dfrac{5}{6} \text{ m/s}^2.$

(ii) The force of 20 kN equals 20 000 N and the mass of 50 t equals 50 000 kg.

$$\text{Using } F = ma$$
$$20\,000 = 50\,000 \times a$$
$$\therefore a = \frac{20\,000}{50\,000} = 0.4.$$

Acceleration $a = 0.4$ m/s².

9.3. GRAVITY AND WEIGHT

Weight has been defined as the force of gravity acting on a body. This force is an example of the universal attraction between masses. Newton's law of gravitation states that 'the force of attraction between any two particles is directly proportional to the product of their masses and inversely proportional to the square of the distance between them.'

Thus the force of attraction $F \propto \dfrac{m_1 m_2}{l^2}$, where m_1 and m_2 are the masses and l is the distance between them and $F = \dfrac{G m_1 m_2}{l^2}$, where G is known as the *universal gravitational constant*.

On the surface of a planet, such as the earth or moon, the force of gravity acting on a body of mass m is given by $\dfrac{G m M}{R^2}$, where M is the mass of the planet and R its radius. But this force of gravity is the weight of the mass m and is equal to mg, where g is the local gravitational acceleration. Hence,

$$mg = \frac{G m M}{R^2},$$

so that $g = \dfrac{G M}{R^2}.$

Values of g and G can be determined by experiment. It is found that all bodies at the same place fall freely with the same value of acceleration, i.e. the 'local gravitational acceleration.' On earth, for most practical purposes, the value of g is taken as 9·81 m/s². However, the value is not the same at all points on the earth's surface and varies by about 0·5 per cent. This is because the earth is not truly spherical. On the surface of the moon the acceleration due to gravity is about one-sixth of the value for the earth (see Example 9.3).

Although it is a constant, G has dimensions and in the SI is equal to $6\cdot67\times10^{-11}$ m³/kg s².

EXAMPLE 9.3. The ratio of the mass of the moon to that of the earth is as 1:81·3. If the ratio of the radius of the moon to the radius of the earth is 0·27:1, what is the value of the acceleration due to gravity on the surface of the moon given that g at the earth's surface is 9·81 m/s²?

SOLUTION

For the moon $\quad g_m = \dfrac{G M_m}{R_m^2}.$

For the earth $\quad g_e = \dfrac{G M_e}{R_e^2}.$

Thus $\quad \dfrac{g_m R_m^2}{M_m} = \dfrac{g_e R_e^2}{M_e}$

and $\quad g_m = \dfrac{g_e R_e^2 M_m}{M_e R_m^2}$

$$= \frac{9\cdot81 \times 1^2 \times 1}{81\cdot3 \times 0\cdot27^2} \text{ m/s}^2$$

$$= 1\cdot65 \text{ m/s}^2.$$

Note: It will be seen that the surface gravity of the moon is approximately one-sixth that of the earth. Thus, a man of mass 70 kg experiences a gravitational force of (weighs) approximately 687 N on earth and only about 115 N on the moon.

9.4. MECHANICAL ENERGY

9.4.1. Potential energy. *Potential energy* is energy of position. The potential energy stored in a body of mass m is equal to the work that may be done by the body if it falls through a vertical distance h.

But work = force × distance,
and in this case the 'force' is the force of gravity acting on the mass, i.e. the weight of the mass.

Thus, potential energy = weight × height
$$= mgh,$$

where m is the mass (kg), g the gravitational acceleration (m/s²), and h the height (m).

9.4.2. Kinetic energy. Kinetic energy is energy of motion. Consider a force F acting on a mass m, which is at rest, and causing a velocity v in a distance s.

Work done $W = Fs$.

But $\quad F = ma$ and $s = \dfrac{v^2 - u^2}{2a}$, where $u = 0$.

Then $\quad W = \dfrac{mv^2}{2} - 0$ and is the kinetic energy

possessed by the body by virtue of its motion.
Thus, kinetic energy = ½ mass × (velocity)²
$$= \frac{mv^2}{2}$$

EXAMPLE 9.4. A locomotive of mass 200 t is travelling at a speed of 54 km/h on a straight and level track. If a braking force of 250 N/t is applied, how far will the locomotive travel before being brought to rest?

SOLUTION

Retarding force = 250 × 200 N
$$= 50\,000 \text{ N}.$$

Kinetic energy of locomotive before brakes applied = ½mv^2
where m = 200 t = 200 000 kg
and v = 54 km/h = 15 m/s.

Therefore, kinetic energy $= \dfrac{200\,000 \times 15^2}{2}$ J
$$= 22\,500\,000 \text{ J}.$$

This kinetic energy equals the work done by the brakes in stopping the locomotive.

But work done = Fs.

$$\therefore Fs = 22\,500\,000$$

$$\text{and } s = \frac{22\,500\,000}{50\,000} = 450.$$

Braking distance = 450 m.

9.5. RESISTED MOTION

The following points should be carefully noted when using the formula $F = ma$.

(i) F is the *accelerating* or *retarding force*. (See Cases I–III below.)

(ii) It is often necessary in working problems to determine the value of acceleration or retardation. This is generally found by using the equations of motion derived in Chapter 8.

9.5.1. Case I. Motion on a horizontal plane. Consider a force P acting on a mass m, against a resistance of R, and producing an acceleration of a, as shown in Fig. 9.1.

The accelerating force $F = (P - R)$ and $F = ma$ is written as $P - R = ma$.

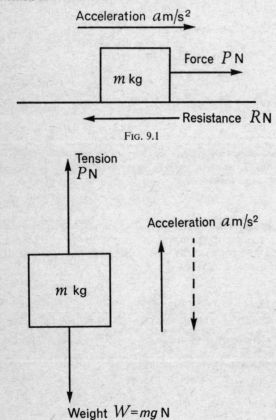

FIG. 9.1

FIG. 9.2

9.5.2. Case II. Vertical motion. Consider a mass m suspended on a cord, and let P be the tension in the cord (see Fig. 9.2).

(a) For motion up or down at constant speed:
acceleration = 0.
Therefore accelerating force $F = 0$.

(b) For motion downwards with acceleration a,
$W \ (= mg)$ is greater than P.
Then the accelerating force $F = (W - P)$,
and $W - P = ma$.

(c) For motion upwards with acceleration a,
P is greater than W.
The accelerating force $F = (P - W)$,
so that $P - W = ma$.

9.5.3. Case III. Motion on a smooth inclined plane. Consider a mass m on an inclined plane where the effects of friction are negligible. This is illustrated in Fig. 9.3.

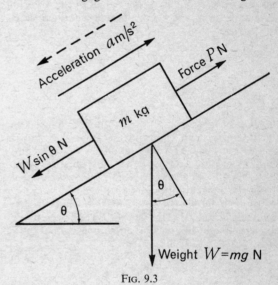

FIG. 9.3

The resolved part of the weight W, acting down the plane $= W \sin \theta$.

(a) If P is greater than $W \sin \theta$, there is motion up the plane with acceleration a.
The accelerating force $F = P - W \sin \theta$, so that
$$P - W \sin \theta = ma.$$

(b) If $W \sin \theta$ is greater than P, there is motion down the plane with acceleration a. Then the accelerating force $F = W \sin \theta - P$.

EXAMPLE 9.5. A planing-machine table has a mass of 400 kg and the frictional resistance at the slides is equivalent to a force of 450 N. If a constant force of 550 N is applied to the table at rest, what speed will it attain after 2 s?

SOLUTION

In this question the value of the acceleration is first obtained by using $F = ma$, where

$$F = (P - R)$$
$$= (550 - 450).$$

Note: $P = 550$ and $R = 450$ (see Case I).

Hence, from $F = ma$,

$$(550 - 450) = 400 \times a$$

$$\text{and } a = \frac{100}{400} = 0.25.$$

To find the velocity after 2 s, the equation $v = u + at$ is used.

Now v is required, $u = 0$, $a = 0.25$, and $t = 2$.
Then $v = 0 + 0.25 \times 2$
$$= 0.5.$$
Velocity after 2 s $= 0.5$ m/s.

EXAMPLE 9.6. A mass of 600 kg has a resistance to motion of 400 N. Calculate the total force in newtons to be applied to the mass to produce a uniform acceleration of 3 m/s².

E.M.E.U.

SOLUTION

Let the total force required be P.
Then, accelerating force $(F) = (P - 400)$
Using $F = ma$,
$$P - 400 = 600 \times 3,$$
$$P = 400 + 1800$$
$$= 2200.$$
Total force required $= 2200$ N.

EXAMPLE 9.7. A passenger lift of mass 1250 kg starts from rest and acquires an upward velocity of 4 m/s in a distance of 10 m. If the acceleration is constant, what is the tension in the lift cable?
Take g as 9.81 m/s².

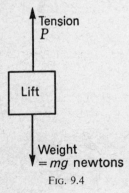

FIG. 9.4

SOLUTION
See Fig. 9.4.
It is first necessary to obtain the acceleration of the lift by using one of the equations of motion.

The known quantities are $u = 0$, $v = 4$, and $s = 10$. It is required to find a.

Using $v^2 = u^2 + 2as$
$$(4)^2 = 0 + 2 \times a \times 10$$
$$16 = 20 a$$
$$\therefore a = \frac{16}{20} = 0.8 \text{ m/s}^2.$$

Let P be the tension in the cable.

Note: If the lift had been stationary, the tension in the cable would have been due only to the gravitational pull (i.e. mg) equal to 1250 kg $\times$ 9.81 m/s² or 12 262.5 N.

Since the lift is accelerating upwards the tension P is greater than 12 262.5 N.

Accelerating force $F = P - 12\ 262.5 = ma$
$$= 1250 \times 0.8$$
$$\therefore P = 12\ 262.5 + 1250 \times 0.8$$
$$= 12\ 262.5 + 1000$$
$$= 13\ 262.5.$$
Tension in the cable $= 13.2625$ kN.

9.6. CIRCULAR OR ANGULAR MOTION

Many problems in engineering are concerned with rotating masses such as flywheels and pulleys. It is therefore necessary to be able to apply to circular motion some of the methods used in linear motion.

One of the fundamental quantities used in circular motion is the *radian (rad)*.

A radian is the angle subtended at the centre of a circle by an arc equal in length to the radius (see Fig. 9.5).

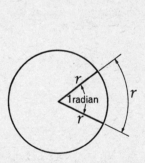
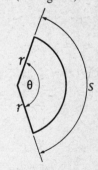

FIG. 9.5. The radian   FIG. 9.6

The relationship between arc length, radius, and angle in radians is illustrated in Fig. 9.6.

Let $s =$ length of arc,
$\theta =$ angle in radians,
$r =$ radius of the circle.

From the definition of a radian,

$$\frac{s}{r} = \theta,$$

$$\therefore s = r\theta,$$

i.e. length of arc $=$ radius $\times \theta$ (in radians).

In the equation $s = r\theta$ if s is equal to the circumference of the circle,

$$\text{then } 2\pi r = r\theta,$$
$$\therefore 2\pi = \theta.$$

That is, 1 *revolution* = 2π *radians* ($= 360°$).

9.6.1. Angular displacement. In linear motion the symbol s represents the distance travelled or linear displacement. The Greek letter θ is the corresponding symbol for displacement in angular motion.

The displacement θ is measured in radians.

9.6.2. Angular velocity. *Angular velocity* may be defined as *the rate of increase of the angle* θ (*in radians*) *with respect to time.*

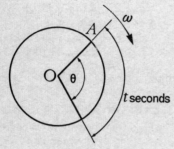

FIG. 9.7. Angular velocity

In Fig. 9.7 the line OA is rotating about the point O and has moved through an angle θ rad in t s.

$$\text{Average angular velocity} = \frac{\text{angle moved through}}{\text{time taken}}$$

$$= \frac{\theta(\text{rad})}{t(\text{s})}$$

$$= \frac{\theta}{t} \, (\text{rad/s})$$

Angular velocity $\dfrac{\theta}{t}$ is given the symbol ω.

$$\text{Hence } \omega = \frac{\theta}{t},$$

where ω is measured in rad/s.

9.6.3. Relationship between angular and linear velocity. Consider the point A on the outer rim of the flywheel in Fig. 9.8.

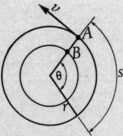

FIG. 9.8. Relationship between angular and linear velocity

Suppose the wheel rotates through θ (rad) in t (s).

Using length of arc = angle in radians $\times$ radius
$$s = r\theta.$$

To obtain the average velocity divide both sides of the equation by t, then

$$\frac{s}{t} = r\frac{\theta}{t},$$

$$\text{but } \frac{s}{t} = v \text{ and } \frac{\theta}{t} = \omega.$$

$$\therefore v = r\omega$$

or linear velocity = angular velocity $\times$ radius.

Note: A point such as B on the inner rim of the flywheel has the same angular velocity as A but a smaller linear velocity.

9.6.4. Units. If the radius r is in metres and the angular velocity ω in rad/s then the linear velocity v is expressed in m/s.

In many problems the angular velocity is expressed in rev/min, and before substituting into the formula $v = r\omega$ the angular velocity must be converted to rad/s or rad/min.

$$N \text{ rev/min} = 2\pi N \text{ rad/min}$$

$$= \frac{2\pi N}{60} \text{ rad/s.}$$

9.6.5. Angular acceleration. A similar relationship to that of linear and angular velocity exists between linear and angular acceleration,

i.e. linear acceleration = radius $\times$ angular acceleration,
$$\text{or } a = r\alpha,$$

where angular acceleration is measured in rad/s².

EXAMPLE 9.8

Convert (a) 50 rev/min to rad/s
(b) 20 rad/s to rev/min.

SOLUTION

$$\text{(a) } N \text{ rev/min} = \frac{2\pi N}{60} \text{ rad/s,}$$

$$\therefore 50 \text{ rev/min} = \frac{2\pi \times 50}{60} \text{ rad/s}$$

$$= 5\cdot2 \text{ rad/s.}$$

(b) 2π radians = 1 revolution.

$$\therefore 20 \text{ rad/s} = \frac{20}{2\pi} \text{ rev/s}$$

$$= \frac{20}{2\pi} \times 60 \text{ rev/min}$$

$$= \frac{600}{\pi} \text{ rev/min}$$

$$= 191 \text{ rev/min.}$$

EXAMPLE 9.9. The wheels of a motor cycle are 0·6 m diameter. If the cycle is travelling at 54 km/h calculate the angular velocity of the wheels in rad/s.

SOLUTION

Convert 54 km/h to m/s.

$$54 \text{ km/h} = \frac{54 \times 1000}{60 \times 60} \text{ m/s}$$
$$= 15 \text{ m/s}.$$

Since the cycle is moving at 15 m/s, the linear velocity of a point at the periphery of the wheel is 15 m/s.

Using $v = r\omega$,
$\quad 15 = 0·3 \times \omega$ (diameter = 0·6 m; radius = 0·3 m),
$\quad \omega = 50$ rad/s.

Angular velocity of wheel = 50 rad/s.

9.7. TORQUE

When a force produces rotation of a body without motion of translation, there must be a second force acting upon the body equal and opposite to the first so that the resultant force is zero.

Consider the case of a spanner being applied to a nut (Fig. 9.9).

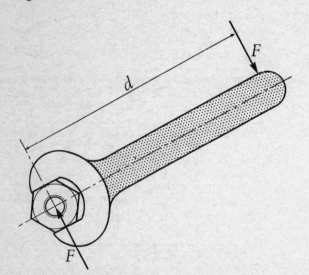

FIG. 9.9. Torque or turning moment

When the hand force F is applied at the end of the spanner an equal and opposite parallel force is set up at the nut.

Note: If this were not so there would be motion of translation in the direction of the force F.

When two equal and opposite forces act on a body they are said to form a couple.

The moment of the couple (Fig. 9.9) = force $F \times$ distance between couple d
$$= F \times d.$$

The moment of a couple is referred to as torque or turning moment.

$$\textbf{Torque} = \textbf{magnitude of force} \times \begin{array}{c}\textbf{length of}\\\textbf{arm of torque.}\end{array}$$

Torque has the same unit as moment of a force, i.e. newton-metre (N m).

9.7.1. Work done by a torque. Consider the work done by a force of F (N) acting at right-angles to an arm OA, of length r (m), and turning through an angle θ (rad) (see Fig. 9.10).

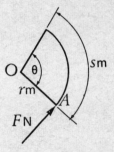

FIG. 9.10. Work done by a torque

$$\begin{aligned}\text{Work done} &= \text{distance moved} \times \text{force}\\ &= s \times F \text{ (N-m or J)}\\ \text{But } s &= r\theta \text{ (m)}\\ \therefore \text{ work done} &= r\theta \times F\\ &= Fr \times \theta\\ \text{But torque } T &= Fr,\\ \therefore \text{ work done} &= T\theta \text{ (N-m or J)}.\end{aligned}$$

$$\begin{array}{c}\textbf{Work done}\\\textbf{by a torque}\end{array} = \textbf{torque} \times \textbf{angle turned through (in radians)}.$$

9.7.2. Power. Suppose the force F (N) in Fig. 9.10 rotates at n (rev/s).

Referring to Chapter 7, the power developed is given by:

$$\text{power} = \frac{\text{work done (J)}}{\text{time taken (s)}} \text{ (J/s or W)}.$$

Work done by a torque = $T\theta$ (J).

Work done per second = $T \times$ angle turned through (rad/s)

$$= T \times 2\pi n \text{ } (n \text{ rev} = 2\pi n \text{ rad})$$
$$\therefore \text{ power} = 2\pi n T \text{ (W)}.$$

EXAMPLE 9.10. The cutting resistance of a lathe tool is 1·2 kN and the average diameter of the work piece is 80 mm. If it is revolving at 2 rev/s calculate:

(a) the work done per minute, and
(b) the power used in the cutting.

SOLUTION

(a) Work done per minute = torque × angle turned through per minute.

$$\text{Torque } T = \text{force} \times \text{torque arm length}$$

$$= 1200 \times 0.04 \text{ N m}$$

(1·2 kN = 1200 N: 40 mm = 0·04 m).

Angle turned through rad/min = $2\pi n \times 60$,

therefore work done/min = $1200 \times 0.04 \times 2\pi \times 2 \times 60$

$$= 36\ 173 \text{ J } (36.173 \text{ kJ}).$$

(b) Power = $\dfrac{\text{work done}}{\text{time taken}} = \dfrac{36\ 173}{60}$ W = 603 W.

EXAMPLE 9.11. A lift cage has a mass of 600 kg and is supported by a cable which is wrapped around a pulley of 1·5 m diameter. When carrying a dead load of 250 kg the cage ascends with uniform acceleration, reaching a speed of 6 m/s after rising 20 m from rest. Calculate

(a) the tension in the cable; and
(b) the torque on the pulley.
(Take $g = 9.81$ m/s²).

SOLUTION

See Fig. 9.11.

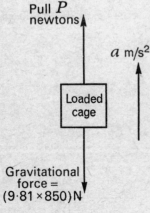

Pull *P* newtons

a m/s²

Loaded cage

Gravitational force = (9·81 × 850)N

FIG. 9.11

(a) Let *P* newtons be the tension (pull) in the cable.
To determine acceleration (*a*),

using $v^2 = u^2 + 2as$,

where $v = 6$ m/s, $u = 0$ and $s = 20$ m.

$$6^2 = 0 + 2a \times 20,$$

$$36 = 40a,$$

$$\therefore \quad a = \frac{36}{40} = 0.9 \text{ m/s}^2.$$

Acceleration = 0·9 m/s².

Since the lift is accelerating upwards the tension in the cable is greater than the load.
Therefore $(P - 9.81 \times 850)$ is the accelerating force *F*.
Using $F = ma$,

$$(P - 9.81 \times 850) = 850 \times 0.9,$$

$$\therefore \quad P - 8339 = 765,$$

$$P = 8339 + 765$$

$$= 9104.$$

Tension in the cable = 9104 N.

(b) Torque = force × torque arm length

$$\left(\text{torque arm length} = \frac{\text{diameter}}{2}\right)$$

$$\therefore \qquad \text{torque} = 9104 \times \frac{1.5}{2} \text{ N m}.$$

$$= 6828 \text{ N m}.$$

Torque on the pulley = 6828 N m.

9.7.3. Work diagrams for torques. It has been shown in section 7.1.1. that an amount of work may be represented by a force-distance diagram drawn to scale. The work done by a constant torque is given by the equation $W = T\theta$ (J), which can be represented by the area of the torque–angular displacement diagram shown in Fig. 9.12. The fact that the area of a torque–angular displacement diagram represents the work done by a torque is particularly useful when the torque is variable as in the case shown by Fig. 9.13. Any convenient method of determining the area of the work diagram may be employed.

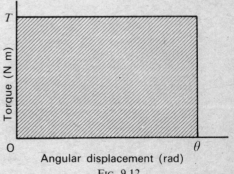

FIG. 9.12

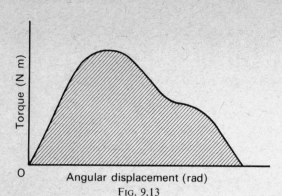

FIG. 9.13

9.8. FRICTIONAL TORQUE

Friction has many advantages and disadvantages. For example, it is an advantage in brakes and clutches and a disadvantage in journal bearings.

When a shaft revolves in a journal bearing a frictional resistance, which is tangential to the shaft, opposes the motion as shown in Fig. 9.14. The shaft exerts a force of mg (N) on the bearing so that the frictional resistance R is given by the equation:

$$R = \mu mg \text{ (N)}.$$

Let the shaft, of diameter d (m), rotate at n (rev/s) which is $2\pi n$ (rad/s). Then the frictional torque T is given by:

$$T = \mu mg \times \frac{d}{2}$$
$$= Rd/2 \text{ (N m)}.$$

The power transmitted by a torque is given by $P = 2\pi nT$ (W) (see section 9.7.2).
∴ the power lost in rotation against friction in a journal bearing is given by

$$P = 2\pi n \times Rd/2$$
$$= \pi nRd \text{ (W)}$$

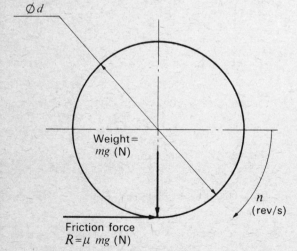

FIG. 9.14

The above equation may also be applied to brakes, clutches, and other applications employing frictional torque. It must be remembered that n is the speed of rotation in revolutions per second, R is the friction force (resistance) in newtons, and d is the effective diameter in metres.

SUMMARY

From Newton's second law, force = mass × acceleration,

$$F = ma.$$

F is the accelerating or retarding force.
Force (N) = mass (kg) × acceleration (m/s²).
A newton is that force which, when acting on a mass of one kilogram produces an acceleration of one metre per second per second.

Circular motion.

Linear velocity = angular velocity × radius,
$$v = r\omega.$$
Linear acceleration = angular acceleration × radius
$$a = r\alpha.$$

ω is measured in rad/s,
α is measured in rad/s².

$$N \text{ rev/min} = \frac{2\pi N}{60} \text{ rad/s}.$$

Torque = magnitude of force × torque arm length.
The unit of torque is the newton-metre (N m).
Work done by torque
= torque × angle turned through in radians.

$$W = T\theta.$$

Power developed (W) = $2\pi nT$, where T is the torque in N m and n is measured in rev/s.

Frictional torque $T = \mu mgd/2$ (N m).

EXERCISES 9

1. What force in newtons will be required to accelerate a mass of 5 kg at 3 m/s²?

Most relevant section: 9.2

U.E.I.

2. A tangential force of 40 N acts at the rim of a 0·2 m diameter wheel and turns it through 10 revolutions. What work is done?

Most relevant section: 9.7.1

U.E.I.

3. A body moves round a circular path of 4 m diameter with a constant angular velocity of 100 rad/s. What

is the magnitude and direction of the linear velocity of the body at any instant?

Most relevant section: 9.6.3

U.E.I.

4. A driving belt moving at 6 m/s is required to turn a pulley at 8 rev/s. What is the diameter of the pulley?

Most relevant section: 9.6.3

Y.C.F.E.

5. Define the units of force (a) the newton and (b) the kilonewton. If a force of 10 N acts on a body to produce an acceleration of 5 m/s², what is the mass of the body?

Most relevant section: 9.2

Y.C.F.E.

6. A car travels at 72 km/h. If the wheel diameter is 460 mm, what is the angular velocity of the wheels in rev/s?

Most relevant section: 9.6.3

Y.C.F.E.

7. In a belt drive a 130 mm diameter pulley rotates at 8 rev/s and drives another pulley 300 mm diameter. Calculate (a) the belt speed in m/s; (b) the speed of the 300 mm diameter pulley.

Most relevant section: 9.6.3

E.M.E.U.

8. Calculate the acceleration produced by
   (a) a force of 20 N acting on a mass of 15 kg,
   (b) a force of 10 kN acting on a mass of 30 t.

Most relevant section: 9.2

E.M.E.U.

9. The wheels of a vehicle are 0·5 m diameter. What is the angular velocity of the wheels in radians per second when the vehicle travels at 48 km/h?

Most relevant section: 9.6.3

C.G.L.I.

10. The force causing an iron core to move into an electric solenoid is 2 N and the mass of the core is 0·3 kg. With what acceleration will it move?

Most relevant section: 9.2

C.G.L.I.

11. (a) A side-and-face cutter of diameter 140 mm has 24 teeth. The surface speed of cutting is to be 22 m/min and the feed is to be 0·25 mm per tooth.
    Calculate, taking π as $\frac{22}{7}$,
    (i) the arbor speed in rev/min, (ii) the feed rate in mm/min.
    (b) When milling a slot the cutting produces a tangential force of 3000 N at the rim of the cutter. Calculate
    (i) the torque in N m, (ii) the power required in kW.

Most relevant sections: 9.6.3, 9.7, and 9.7.2

C.G.L.I.

12. A car of mass 1·2 t travelling at 54 km/h along a level road is brought to rest in 55 m. Determine
    (a) the time taken to come to rest, and
    (b) the average retarding force.

Most relevant sections: 8.2 and 9.2

U.E.I.

13. A vehicle of mass 1 t is brought to rest in 12 m with a retardation of 1·5 m/s². Calculate the average value of the force required to stop it and the work done.
    If all the work done in stopping the vehicle is equally shared between four brakes determine the force and torque on one of the brake drums. Each brake drum has a diameter of 0·3 m. The circumference of the wheels is 2·4 m and there is no slip.

Most relevant sections: 9.2, 7.1, and 9.7

Y.C.F.E.

14. A screw is used to raise a mass of 2 t. The mass is lifted 12 mm in one revolution of the screw. If an effort of 300 N is required at the end of a 450 mm arm fastened to the screw, determine:
    (a) the torque required to turn the screw,
    (b) the work done in turning the arm through 10 revolutions; and
    (c) the efficiency of the lifting operation.
    Take g as 9·81 m/s².

Most relevant sections: 9.7, 9.7.1, and 7.4

Y.C.F.E.

15. A mass of 180 kg is lifted vertically at a steady speed of 0·6 m/s by a chain, whose weight may be neglected, wrapped round a drum whose effective diameter is 450 mm. The drum is driven by an electric motor through a reduction gearing having a ratio of 10/1 and mechanical efficiency of 80 per cent. Calculate:
    (a) the work done on the load in J/s,
    (b) the power output of the motor,
    (c) the drum speed in rev/min,
    (d) the motor speed in rev/min.

Most relevant sections: **7.1, 7.3,** and **9.6.3**

E.M.E.U.

16. The leadscrew of a lathe has a single start thread with a pitch of 6 mm. If the torque to turn the leadscrew and so move the saddle is 0·42 N m find the work done in joules to turn the leadscrew through 10 revolutions.
    If the force required to move the saddle along the bed of the lathe is 220 N find the work done in joules in moving the saddle.
    What is the mechanical efficiency of the drive?

Most relevant sections: **9.7.1** and **7.4**

E.M.E.U.

17. A screwjack has a single start thread of pitch 5 mm. It is used to lift a mass of 0·5 t through 240 mm, efficiency at this load being 55 per cent. Find
    (a) the number of turns made by the screw,
    (b) the work input to the jack, and
    (c) the torque required to turn the screw.
    Take g as 9·81 m/s².

Most relevant sections: **7.1** and **9.7**

C.G.L.I.

18. A constant tractive effort is applied, for a period of 5 seconds, to a vehicle of mass 1000 kg so that it is accelerated from rest to a speed of 15 m/s, against opposing frictional forces equal to 100 N. Determine
    (a) the deceleration on removal of the tractive effort,
    (b) the total time for which the vehicle is in motion, and
    (c) the total distance travelled by the vehicle.

Most relevant sections: **9.2** and **8.2**

U.E.I.

19. (a) With oil-lubricated plain journal bearings an approximate relationship between the coefficient of friction μ, the bearing pressure p N/mm² (load divided by plan area) and the surface velocity of the shaft v m/min is

$$\mu = \frac{0·0032\ \sqrt{v}}{p}$$

Find μ, when $v = 144$ and $p = 0·8$.

(b) In one particular application the coefficient of friction was 0·02, the shaft diameter was 100 mm, and the load on the bearing was 12 000 N. Calculate
(i) the frictional torque in newton-metres,
(ii) the power consumed by friction, in watts, when the shaft rotates at 315 rev/min.

Most relevant section: **9.8**

C.G.L.I.

20. (a) A workpiece, of mean diameter 70 mm, is being turned in a lathe. The spindle speed is 240 rev/min, the area of the chip section is 5 mm².

Calculate

(i) the cutting speed in m/min, (ii) the rate at which material is being removed in mm³/s.
(b) If it takes 1·25 J of energy to remove 1 mm³ of material, calculate
(i) The power in kilowatts consumed at the tool point,
(ii) the torque carried by the lathe spindle, in newton-metres.

Most relevant sections: **9.6.3, 9.7,** and **9.7.2**

C.G.L.I.

21. A horizontal length of shafting exerts a vertically downward force of 4000 N on a plain bearing.
    (a) If the diameter of the shaft is 80 mm and the coefficient of friction is 0·05, calculate the frictional torque in newton-metres.
    (b) The shaft rotates at 315 rev/min. Taking π as $\frac{22}{7}$, (i) convert this speed to an equivalent in radians per second, (ii) calculate, as a fraction of a kilowatt, the power consumed in friction.
    (c) Name one lubricant in each case where at *normal* room temperature the lubricant is, (i) liquid in form, (ii) solid in form.

Most relevant sections: **9.8** and **9.6.4**

C.G.L.I.

22. The torque T on a shaft varies during successive revolutions as shown in the table below:

T (N m)	0	50	87	100	87	50	0
θ (rad)	0	$\frac{\pi}{6}$	$\frac{\pi}{3}$	$\frac{\pi}{2}$	$\frac{2\pi}{3}$	$\frac{5\pi}{6}$	π

Plot these values and estimate from the resulting work diagram,

(a) the average torque, and

(b) the work done per revolution.

Most relevant section: 9.7.3

U.E.I.

23. (a) What is meant by the 'efficiency' of a machine?
(b) A lift cage has a mass of 560 kg and is supported by a cable which is wrapped around a pulley of 1 m effective diameter. When carrying a load of mass 440 kg the cage commences its ascent with a uniform acceleration of 1·5 m/s². Calculate:

(i) the tension in the cable due to the cage and its load,

and

(ii) the driving torque of the pulley.

Take the acceleration due to gravity g as 9·81 m/s².

Most relevant sections: 7.4, 8.2, 9.5.2, and 9.7

U.E.I.

ANSWERS TO EXERCISES 9

1. 15 N. **2.** 251·4 J. **3.** 200 m/s. **4.** 0·24 m. **5.** 2 kg. **6.** 13·8 rev/s. **7.** (a) 3·27 m/s, (b) 3·5 rev/s. **8.** (a) 1·33 m/s², (b) 0·33 m/s². **9.** 53·3 rad/s. **10.** 6·67 m/s². **11.** (a) 50 rev/min, 300 mm/min, (b) 210 Nm, 1·098 kW. **12.** (a) 7·33 s, (b) 2·45 kN. **13.** 1·5 kN, 18 kJ, 955·4 N, 286·6 Nm. **14.** (a) 135 Nm, (b) 8·48 kJ, (c) 28 per cent. **15.** (a) 1059 J/s, (b) 1·32 kW, (c) 25·5 rev/min, (d) 318·7 rev/min. **16.** 26·4 J, 13·2 J, 50 per cent. **17.** (a) 48, (b) 2·14 kJ, (c) 7 N m. **18.** (a) 0·1 m/s², (b) 155 s, (c) 1162·5 m. **19.** (a) 0·048, (b) 12 N m, 395 W. **20.** (a) 52·8 m/min, 4400 mm³/s, (b) 5·5 kW, 218·8 N m. **21.** (a) 8 N m, (b) 33 rad/s, 0·264 kW. **22.** (a) 64·3 N m, (b) 404 J. **23.** (b) 11·31 kN, 5655 N m.

10 Simple machines

10.1. MACHINES

A *machine* may be defined as a device for doing work. In general, a force is applied at one point on a machine in order to overcome a resistance at another point. The applied force known as the effort, is said to be on the input side of the machine while the resistance to be overcome, known as the load, is on the output side. Examples of simple machines are screwjacks, winches, and pulley blocks. In any machine some of the work, or power, input is used to overcome various internal losses such as friction and windage and, as a result, the work, or power, output is somewhat less than the input.

Fig. 10.1 represents any machine in which:

work output = work input−losses.

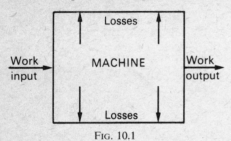

FIG. 10.1

From the foregoing it would appear that, to be useful, a machine must have low losses and a high efficiency. This is often not the case. Many machines, such as the screwjack, are often designed to have a low efficiency so that a small effort can be used to raise, or overcome, a large load.

10.2 LAW OF A MACHINE

The basic characteristic of a machine is the variation of effort with load. A graph of effort against load is a straight line and its equation is of the type $y = ax+b$, where a and b are constants. For an ideal machine, in which there would be no friction and is therefore theoretical, the relationship between effort and load is shown by graph (i) in Fig. 10.2. In a real machine friction is always present, even under no-load conditions, and the relationship between effort and load is illustrated by graph (ii). It will be observed from the diverging graphs that in a real machine the frictional loss increases with increase in load. For a real machine the equation of the graph is known as *the law of the machine* and is:

$$F_{in} = aF_{out}+b$$

The constant a is the slope of the straight-line graph and the constant b, which is the point of intersection on the y-axis, represents the effort required to overcome friction at no-load.

10.3 MECHANICAL ADVANTAGE

The *mechanical advantage* (abbreviated M.A.) of a machine is defined as the ratio of the load to the effort. Both the load and the effort must be measured in the same units, e.g. newtons, and therefore the mechanical advantage of a machine will have no units. The mechanical advantage increases with the load but is not proportional to it as is shown by Fig. 10.3.

$$\text{M.A.} = \frac{\text{load } F_{out} \text{ (N)}}{\text{effort } F_{in} \text{ (N)}}.$$

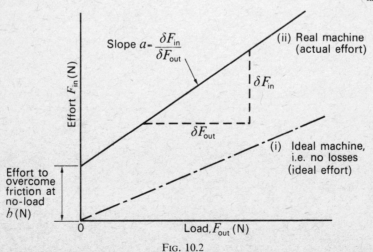

FIG. 10.2

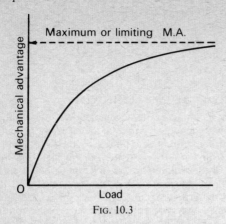

Fig. 10.3

The law of a machine being $F_{in} = aF_{out} + b$, then

$$\text{M.A.} = \frac{F_{out}}{aF_{out} + b} = \frac{1}{a + \dfrac{b}{F_{out}}}.$$

When the load F_{out} is very large the term b/F_{out} is negligible and the maximum or limiting value of the mechanical advantage is given by

$$\text{M.A.}_{lim} = \frac{1}{a}.$$

10.4. VELOCITY RATIO

The *velocity ratio* (abbreviated V.R.) of a machine is defined as the ratio of the velocity of the effort to the velocity of the load. However, by definition, velocity is the rate of change of distance with time, and since both the effort and load move through the same period of time, it is often more convenient to express the velocity ratio as:

$$\text{V.R.} = \frac{\text{distance moved by effort}}{\text{distance moved by load}}.$$

The velocity ratio of a machine depends upon its geometry (i.e. its design and construction) and provided that this remains unaltered, is a constant.

Note: velocity ratio has no units.

10.5. EFFICIENCY

The *efficiency* of any machine (or process) is defined as the ratio of the power output to the power input. The power input to a machine must be sufficient to provide the necessary power output and to overcome the various losses, e.g. friction, which may occur in the machine itself.

$$\text{Efficiency } \eta = \frac{\text{power (or work in a given time) output}}{\text{power (or work in the same time) input}}.$$

Since work done = distance moved × force (in direction of motion),

$$\text{then} \qquad \eta = \frac{\text{distance moved by load} \times \text{load}}{\text{distance moved by effort} \times \text{effort}}$$

$$= \frac{1}{\text{V.R.}} \times \text{M.A.}$$

$$\therefore \qquad \eta = \frac{\text{M.A.}}{\text{V.R.}}$$

The velocity ratio is, in general, a constant for a given machine so that the efficiency will increase with the load, as illustrated by Fig. 10.4, in the same way as the mechanical advantage. When the mechanical advantage has reached its limiting value of $1/a$, then the efficiency will also reach its maximum or limiting value. Thus, from $\eta = \text{M.A.}/\text{V.R.}$, limiting efficiency,

$$\eta_{lim} = \frac{1}{a(\text{V.R.})}.$$

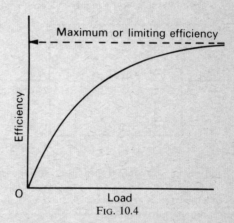

Fig. 10.4

10.6. REVERSAL AND NON-REVERSAL OF A MACHINE

It has been seen that the efficiency of a machine is given by:

$$\eta = \frac{\text{work output}}{\text{work input}}$$

$$= \frac{\text{distance moved by load} \times \text{load}}{\text{distance moved by effort} \times \text{effort}}$$

$$= \frac{s_{out} F_{out}}{s_{in} F_{in}}.$$

Also,

$$\text{work output} = \text{work input} - \text{losses}.$$

It follows that the losses taking place in a machine will be:

$$\text{losses} = \text{work input} - \text{work output}$$

$$= s_{in} F_{in} - s_{out} F_{out}.$$

Now if the effort is removed the machine may tend to reverse due to the load acting as an effort in the opposite direction. Reversal will only take place if $s_{out}F_{out}$ is greater than $s_{in}F_{in} - s_{out}F_{out}$,

i.e. when $s_{out}F_{out} > s_{in}F_{in} - s_{out}F_{out}$,

$$\frac{s_{out}F_{out}}{s_{in}F_{in}} > 1 - \frac{s_{out}F_{out}}{s_{in}F_{in}},$$

$$\eta > 1 - \eta,$$

which is clearly when the efficiency is greater than 0·5 p.u. or 50 per cent. This reversal of a machine, which may take place on removal of the effort, is known as *over-hauling*. For safety reasons, a machine with an efficiency greater than 50 per cent should be fitted with a device, such as a ratchet and pawl, to prevent run-back. A machine with an efficiency of less than 50 per cent cannot overhaul and such a safety device is unnecessary.

EXAMPLE 10.1. A screwjack has a single-start thread of 12 mm pitch. The effort is applied tangentially on a lever at a radius of 350 mm from the centre of the screw. Calculate:

(a) the velocity ratio of the screwjack, and
(b) the load that can be lifted by an effort of 25 N if the efficiency is 30 per cent.

SOLUTION

A screwjack is shown in Fig. 10.13 (p. 89).
(a) Consider 1 revolution of the lever:
effort moves a distance of $2\pi \times 350 = 2200$ mm and the load is raised 12 mm.

$$\therefore \text{ V.R.} = \frac{2200}{12} = 183\tfrac{1}{3}.$$

(b)

$$\text{Efficiency} = \frac{\text{M.A.}}{\text{V.R.}}$$

$$\therefore \qquad \text{M.A.} = \eta \,(\text{V.R.})$$
$$= 0·3 \times 183\tfrac{1}{3}$$
$$= 55.$$

$$\text{But M.A.} = \frac{F_{out}}{F_{in}}$$

$$\therefore \qquad \text{load, } F_{out} = \text{M.A.} \times F_{in}$$
$$= 55 \times 25$$
$$= 1375 \text{ (N)}.$$
$$\text{Load lifted} = 1·375 \text{ kN}.$$

EXAMPLE 10.2. In a test on a certain lifting machine, having a velocity ratio of 50, it was found that an effort of 15 N lifted a load of 200 N, and an effort of 20 N lifted a load of 400 N. Find:

(a) the law of the machine;
(b) the limiting mechanical advantage;
(c) the limiting efficiency, and
(d) the applied force, mechanical advantage, and efficiency when the load is 1·8 kN.

SOLUTION

(a) The law is of the type $F_{in} = aF_{out} + b$, so that when $F_{in} = 15$ and $F_{out} = 200$, then

$$15 = 200a + b, \tag{1}$$

and when $F_{in} = 20$ and $F_{out} = 400$, then

$$20 = 400a + b \tag{2}$$

The pair of simultaneous equations are now solved for a and b.

$$20 = 400a + b \tag{2}$$
$$\underline{15 = 200a + b} \tag{1}$$
$$5 = 200a$$

(by subtraction)

$$\therefore a = \frac{5}{200} = \frac{1}{40}.$$

By substitution using eqn (2):

$$20 = \frac{400}{40} + b$$
$$= 10 + b$$
$$\therefore \qquad b = 20 - 10 = 10.$$

The law of the machine is:

$$F_{in} = \frac{F_{out}}{40} + 10.$$

(b) The limiting M.A. is:

$$\text{M.A.}_{\text{lim}} = \frac{1}{a} = \frac{1}{1/40} = 40.$$

(c) The limiting efficiency is:

$$\eta_{\text{lim}} = \frac{1}{a(\text{V.R.})} = \frac{1}{\dfrac{1}{40} \times 50}$$

$$= \frac{40}{50} = 0·8 \text{ p.u. (80 per cent)}.$$

(d) When the load is 1·8 kN (1800 N):

$$F_{in} = \frac{1800}{40} + 10$$
$$= 45 + 10 = 55.$$

Applied force = 55 N.

$$\text{M.A.} = \frac{F_{out}}{F_{in}} = \frac{1800}{55} = 32·7.$$

$$\eta = \frac{\text{M.A.}}{\text{V.R.}} = \frac{32·7}{50} = 0·654 \text{ p.u. (65·4 per cent)}.$$

10.7. BELT DRIVES

Power is frequently transmitted from one rotating shaft or pulley to another by means of a belt passing round both. Belt drives are more flexible than those obtained by the use of gearing and, in the event of sudden stoppage, slipping of the belt on the shaft or pulley reduces the possibility of excessive damage.

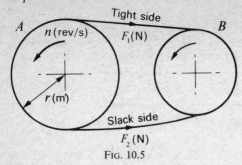

FIG. 10.5

In Fig. 10.5 two pulleys A and B are connected by an open belt and rotate in the same direction as shown. Initially when the shafts are at rest the tension throughout the belt is the same. When power is applied to the driver shaft A the tensions in the two sides of the belt are no longer equal. The tension on the tight or driving side F_1 is greater than that on the slack side F_2.

The difference between the tensions F_1 and F_2 is the effective driving force causing rotation of the driven pulley. To determine the power transmitted by the driver A,

torque due to $F_1 = F_1 r$ (N m),
torque due to $F_2 = F_2 r$ (N m),
$\therefore$ resultant torque $= F_1 r - F_2 r = (F_1 - F_2)r$ (N m).

If the pulley A rotates at n (rev/s),
work done per second $= (F_1 - F_2)r \times 2\pi n$ (J/s)

(where $(F_1 - F_2)r =$ torque T).
$\therefore$ power $= (F_1 - F_2)r \times 2\pi n$ (W).

The above expression can also be written in terms of the radius and rotational speed of the driven pulley B.

Linear velocity of the belt, $v = 2\pi r n$; the power may therefore be expressed as

$$\text{power} = (F_1 - F_2)v \text{ (W)}$$

where v is expressed in m/s.

In Fig. 10.6 the pulleys A and B are connected by a crossed belt which causes the driven pulley B to rotate in the opposite direction to pulley A.

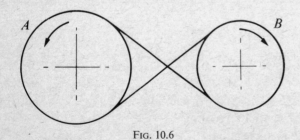

FIG. 10.6

Power transmission by means of belts and pulleys is generally carried out using flat-belts or vee-belts and in each case some belt-slip and belt creep takes place resulting in a loss of power. The tension ratio F_1/F_2 for a

flat-belt is about 1·5–3 and for a vee-belt is about 6–9, which means that a vee-belt drive can transmit about 3–4 times more power than a flat-belt drive under similar conditions. For this reason the vee-belt has largely replaced the flat-belt.

Assuming that no slip or creep occurs between a belt and the pulleys then any point on the outer rim of the pulleys will have the same linear velocity as the belt.

Let $d_A =$ the diameter of pulley A,
   $d_B =$ the diameter of pulley B,
   $n_A =$ the rotational speed of pulley A, and
   $n_B =$ the rotational speed of pulley B.

Then the circumference of pulley $A = \pi d_A$,

the circumference of pulley $B = \pi d_B$
the linear velocity of a point on the outer rim of pulley A $\Big\} = \pi d_A n_A$, and
the linear velocity of a point on the outer rim of pulley B $\Big\} = \pi d_B n_B$.

Thus,

$$\pi d_A n_A = \pi d_B n_B$$

and
$$\frac{n_A}{n_B} = \frac{d_B}{d_A},$$

which shows that the ratio of the pulley rotational speeds is the inverse of the ratio of their diameters.

EXAMPLE 10.3. The tensions in the tight and slack sides of a driving belt are 1000 N and 460 N respectively. If the driving pulley has a diameter of 0·6 m and rotates at 5 rev/s, find
(a) the driving torque, and
(b) the power transmitted.

SOLUTION

(a) Resultant force = difference between tensions
              $= 1000 - 460$ N
              $= 540$ N.
   Torque = resultant force × torque arm length
         $= 540$ N × 0·3 m
         $= 162$ N m.
Driving torque = 162 N m.

(b) Work done per second = torque × angle (rad) turned through per second
              $= 162$ N m × 2π × 5 rad/s
              $= 5087$ J/s.

Power = work done per second = 5087 W or 5·087 kW.
Power transmitted = 5·087 kW.

10.8. GEAR DRIVES

10.8.1. Gears and simple gear-trains. Consider two pulleys A and B to be in contact as shown in Fig. 10.7. If the *driver* pulley A is made to rotate in an anticlockwise direction, then the *driven* pulley B will rotate in the opposite direction, i.e. clockwise. Such an arrangement could only be used for very light loads as it relies entirely upon the friction between the pulleys to transmit power from *A* to *B*.

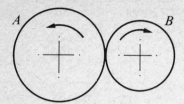

FIG. 10.7

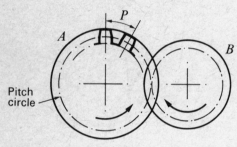

FIG. 10.8

For positive and accurate power transmission from one shaft to another, toothed gear-wheels are used as shown in Fig. 10.8, the teeth being equally spaced around each wheel. For a given wheel the distance between the centre-lines of adjacent teeth, measured around the pitch circle, is known as its circular pitch *p*. The number of teeth on each of the wheels will be proportional to their *pitch circle diameters* (p.c.d.).

Let T_A = the number of teeth on wheel *A* = $\pi d_A / p$,

T_B = the number of teeth on wheel *B* = $\pi d_B / p$,

n_A = the rotational speed of wheel *A*, and

n_B = the rotational speed of wheel *B*.

In unit time wheel *A* advances by $T_A n_A$ teeth and wheel *B* by $T_B n_B$ teeth.

$$\therefore \qquad T_A n_A = T_B n_B$$

and

$$\frac{n_A}{n_B} = \frac{T_B}{T_A},$$

which shows that the rotational speeds of meshing gear-wheels are in inverse ratio to the numbers of teeth on the wheels (and to their diameters).

The p.c.d. of a gear-wheel is given by

$$\textbf{p.c.d.} = \frac{\textbf{number of teeth} \times \textbf{circular pitch}}{\pi}$$

$$= \frac{Tp}{\pi},$$ where p/π is often referred to as the *module M*.

$$\therefore \qquad \text{p.c.d.} = TM.$$

Where two gear-wheels, *A* and *B*, are required to rotate in the same direction, an *idler wheel* is placed between them as shown in Fig. 10.9. The idler does not affect the rotational speed of wheel *B* so that $n_A / n_B = T_B / T_A$. Idlers are also used to take up space between the driving and driven gear-wheels, *A* and *B*, when their shafts are too far apart. The arrangements shown in Figs 10.8 and 10.9 are examples of simple gear-trains in which all the gear-wheels lie in the same plane.

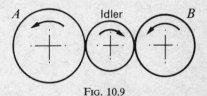

FIG. 10.9

10.8.2. Compound gear-trains. The arrangement shown in Fig. 10.10 is a compound gear-train in which the gear-wheels do not all lie in the same plane. The driver wheel *A*, which is attached to the input shaft, is connected to the driven wheel *B*, attached to the output shaft, by means of the *compound gear-wheel* $C_1 C_2$. Both parts of the compound gear $C_1 C_2$ move at the same rotational speed, being rigidly fixed to one another.

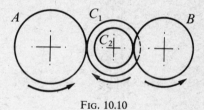

FIG. 10.10

Consider section AC_1 as a simple gear-train with wheel *A* driving wheel C_1.

Then

$$\frac{n_A}{n_{C_1}} = \frac{T_{C_1}}{T_A}$$

$$\therefore \qquad n_{C_1} = \frac{n_A T_A}{T_{C_1}},$$

Now consider section $C_2 B$ as a second simple gear-train with wheel C_2 driving wheel *B*. Since C_1 and C_2 are connected and rotate at the same speed n_{C_1}

then

$$\frac{n_{C_1}}{n_B} = \frac{T_B}{T_{C_2}}$$

$$\therefore \qquad n_{C_1} = \frac{n_B T_B}{T_{C_2}}.$$

Thus,

$$\frac{n_A T_A}{T_{C_1}} = \frac{n_B T_B}{T_{C_2}}$$

and

$$\frac{n_A}{n_B} = \frac{T_B T_{C_1}}{T_A T_{C_2}},$$

i.e. $\dfrac{\text{input rotational speed}}{\text{output rotational speed}} =$

$$\frac{\text{product of teeth on driven wheels}}{\text{product of teeth on driving wheels}}.$$

10.9. PULLEY BLOCK AND TACKLE

The pulley block and tackle is a simple lifting device which is commonly used in cranes, lifts, etc. Fig. 10.11 shows a pulley system in which there are two pulleys in each block and, therefore, four sections of rope between the pulleys which support the load.

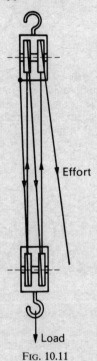

FIG. 10.11

Suppose the effort to move a distance of 1 m, then each of the four parts of the rope supporting the load would shorten by $\frac{1}{4}$ m. Thus the velocity ratio of the pulley system shown in Fig. 10.11 is $1/\frac{1}{4}$, which is 4. It will be observed that there are four pulleys and four sections of rope between the pulleys. It follows that the velocity ratio of any such pulley block and tackle is given by:

V.R. = number of pulleys,

and

V.R. = number of sections of rope supporting the load.

10.10. WESTON DIFFERENTIAL PULLEY BLOCK

In the Weston differential pulley block the upper and lower pulleys are generally connected by an endless chain as shown in Fig. 10.12. The links of the chain fit into recesses around the circumference of the pulleys. The upper pulleys, which are fixed together, have different diameters and hence a different number of recesses.

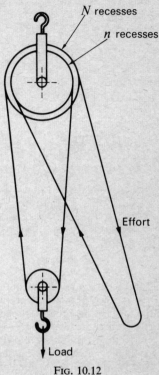

FIG. 10.12

Suppose the load to be raised and the upper pulleys to rotate one revolution. The effort chain will move N links, while at the same time the load chain will wind on to the large pulley by N links but unwind from the small pulley by n links. The net shortening of the load chain therefore, is, $N-n$ links, and the load, which is supported by two sections of the chain, is lifted a distance equal to the length of $(N-n)/2$ links. Thus the velocity ratio of a Weston differential pulley block is given by

$$\text{V.R.} = \frac{N}{(N-n)/2} = \frac{2N}{N-n}.$$

10.11. SCREWJACK

The simple screwjack shown in Fig. 10.13 consists essentially of a vertical screw having n starts of pitch p (mm), and hence a lead of np (mm). The load is carried axially by the screw and the effort is applied tangentially to a handle.

Let the screw rotate one revolution. The effort will have moved $2\pi r$ (mm), while the load will have moved

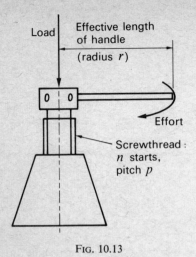

FIG. 10.13

np (mm). Then the velocity ratio of the screwjack is given by:

$$\text{V.R.} = \frac{2\pi r}{np}.$$

10.12. HYDRAULIC JACK

A cross-section of a simple form of hydraulic jack is shown in Fig. 10.14. A cylinder and piston of large diameter is connected by a pipe to a pump of small diameter. Now when a liquid, e.g. oil, completely fills a closed container and a pressure is applied to the liquid at any point, the pressure is carried throughout the whole of the liquid. Oil from a reservoir (not shown) is pumped into the cylinder by means of a handle and the load piston is forced to move out. When the load is to be lowered, a

valve (not shown) is operated which allows the oil to return to the reservoir.

Let the pump piston, of diameter d, move downwards by a distance s_{in} so that the volume of oil displaced is $\pi d^2 s_{\text{in}}/4$. The load piston, of diameter D, will be pushed upwards through a distance s_{out}, the volume moved through being $\pi D^2 s_{\text{out}}/4$ which is equal to the volume displaced by the pump. Then

$$\frac{\pi d^2 s_{\text{in}}}{4} = \frac{\pi D^2 s_{\text{out}}}{4}$$

and $s_{\text{in}}/s_{\text{out}} = D^2/d^2$ which means that for the hydraulic section of the jack the velocity ratio is given by:

$$\text{V.R.} = D^2/d^2.$$

Note: Should the total velocity ratio of the jack be required then the V.R. of the operating handle must also be taken into account.

10.13. TOGGLE-JOINT MECHANISM

The principle of the toggle-joint mechanism is shown in Fig. 10.15. This simple but very important link mechanism is used in applications such as presses of various types where only a short movement of the effort F_{in} is required. The thrust F_{out} at the end of the toggle-arms is given by

$$F_{\text{out}} = \frac{F_{\text{in}}}{2 \tan \theta}.$$

The velocity ratio of this device is *variable* which will be understood after a careful inspection of Fig. 10.16.

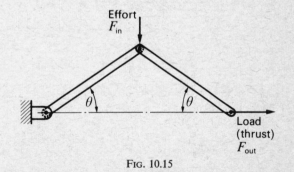

FIG. 10.15

10.14. WORM AND WHEEL

The construction of the worm and wheel is shown in Fig. 10.17. The worm is a screw having n starts of pitch p mm and hence a lead of np mm. The worm engages with a gear wheel having T teeth of circular pitch p mm, and this gear wheel is attached to the load drum having a diameter d mm. An effort is applied to a pulley, of diameter D mm, which is fixed on the worm shaft.

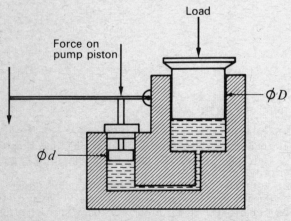

FIG. 10.14

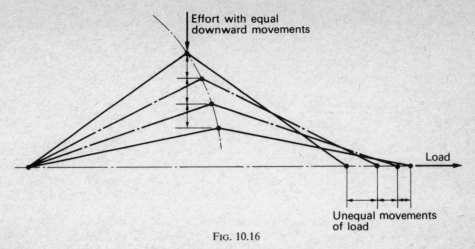

FIG. 10.16

Suppose the worm to rotate through one revolution so that the effort moves a distance of πD mm. The gear-wheel advances by n teeth and the distance moved by the load is $\pi dn/T$ mm. The velocity ratio is thus given by

$$\text{V.R.} = \frac{\pi D}{\pi dn/T} = \frac{DT}{dn}.$$

same time the rack will move a distance of Tp (mm), so that the velocity ratio of the rack and pinion is given by:

$$\text{V.R.} = \frac{2\pi r}{Tp}.$$

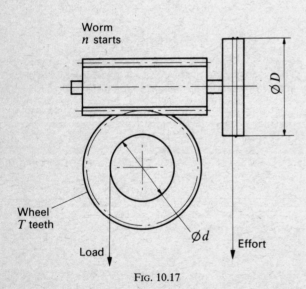

FIG. 10.17

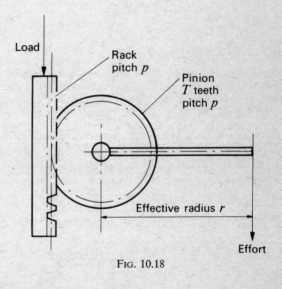

FIG. 10.18

10.15. RACK AND PINION

In the rack-and-pinion mechanism, as shown in Fig. 10.18, the circular pitch, p mm, of the pinion must equal the linear pitch of the rack. The effort is applied at an effective radius r (mm), so that for one revolution of the pinion the effort moves a distance of $2\pi r$ (mm). At the

10.16. GEARED WINCH

The principle of the geared winch is shown in Fig. 10.19. The reduction ratio G may be obtained using either simple or compound gear-trains, the latter leading to a larger velocity ratio.

Let the effort be applied at an effective radius R (mm), and make one revolution thus moving a distance of $2\pi R$ (mm). The load drum will turn through $1/G$ revolution and the load will move a distance of

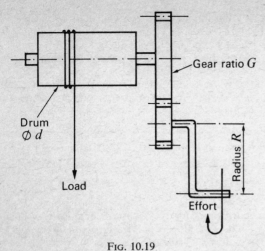

Fig. 10.19

$\pi d/G$ (mm). The velocity ratio of the winch therefore, is, given by

$$\text{V.R.} = \frac{2\pi R}{\pi d/G}$$

$$= \frac{2GR}{d}.$$

SUMMARY

In any machine: work output = work input−losses.
The law of a machine is: Load = a.effort+b, where a and b are constants.
The mechanical advantage is:

$$\text{M.A.} = \text{load/effort},$$
$$\text{M.A.}_{11m} = 1/a.$$

The velocity ratio is:

$$\text{V.R.} = \frac{\text{effort distance}}{\text{load distance}}.$$

The efficiency of a machine is:

$$\eta = \frac{\text{output}}{\text{input}} = \frac{\text{M.A.}}{\text{V.R.}},$$

$$\eta_{11m} = \frac{1}{a(\text{V.R.})}.$$

A machine will not run-back (overhaul) provided that its efficiency is less than 50 per cent.

EXERCISES 10

1. (a) The table assembly of a milling machine is raised by operating a handle, the effort being applied tangentially at a radius of 210 mm. One turn of the handle raises the table by 5 mm. Calculate the velocity ratio of the mechanism.
(b) The table assembly exerts a force of 1320 N. If a steady effort of 20 N applied at the handle lifts the table at constant velocity, calculate, for this particular loading,
(i) the mechanical advantage,
(ii) the efficiency, as a percentage.
(c) State why the table assembly will not run back when the effort is removed.

Most relevant sections: 10.4, 10.3, 10.5, and 10.6

C.G.L.I.

2. (a) A bolt is screwed with a metric thread designated M8×1. This thread has a pitch of 1 mm. A turning force is provided at the end of a single-ended spanner at a radius of 210 mm. Calculate the velocity ratio of the operation. (Take π as $\frac{22}{7}$.)
(b) Assuming the root diameter of the thread to be 7 mm, and the failing tensile stress of the bolt material to be 400 N/mm² or MPa, calculate the tensile load, in newtons, that would cause the bolt to fail in tension.
(c) Calculate the effort needed, in newtons, at the end of the spanner to cause the bolt to fail in tension assuming an efficiency of 20 per cent.

Most relevant sections: 10.4, 5.3, and 10.5

C.G.L.I.

3. During a test on a lifting machine it was found that an effort of 20 N was required to raise a load of 250 N and an effort of 50 N was required to raise a load of 750 N. The law of the machine is of the form.

$$E = aW+b,$$

where E is the effort input force, W is the load output force, and a and b are constants.
(a) Determine the values of the constants a and b and state the law of the machine.
(b) Use this law to calculate the load that can be lifted by an effort of 80 N.
(c) Find the efficiency of the machine at this load, if the velocity ratio of the machine is 40.

Most relevant sections: 10.2 and 10.5

U.E.I.

4. The tensions in the tight and slack sides of a driving belt are 1000 N and 460 N respectively. If the driving pulley has a diameter of 400 mm and rotates at 300 rev/min, find:
(a) the driving torque,
(b) the power transmitted, and
(c) the linear speed of the belt.

Most relevant sections: 10.7 and 9.6.3

U.E.I.

5. Fig. 10.20 shows the gear-drive for a machine table. The motor transmits 18 kW at 28 rev/s. Calculate:
(a) the speed of gear A in rev/s,
(b) the velocity of the machine table, in m/s, if the pitch circle diameter of gear A is 640 mm,
(c) the maximum torque available on the shaft of gear A if the efficiency of the gear-drive is 80 per cent,
(d) the maximum force available to drive the machine table.

***

Most relevant sections: 10.8.2, 9.6.3, and 9.7

***

U.E.I.

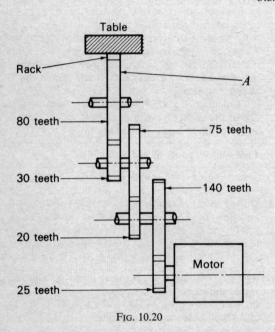

Table

Rack

A

80 teeth

75 teeth

30 teeth

140 teeth

20 teeth

Motor

25 teeth

FIG. 10.20

6. A certain pulley block and tackle has four pulleys in both the upper and lower blocks. The tackle is used to lift a mass of 100 kg with an effort of 150 N. Take the local gravitational acceleration g as 9·81 m/s² and find:
(a) the velocity ratio,
(b) the mechanical advantage, and
(c) the efficiency.

***

Most relevant sections: 10.9, 10.3, and 10.5

***

7. (a) A Weston differential pulley block is chain operated and has 20 flats on the large pulley and 19 flats on the small pulley. Calculate the velocity ratio.

(b) It is found that this block will lift a load of 1·1 kN with an effort of 100 N and a load of 1·9 kN with an effort of 140 N. If the load W and effort E are connected by a law of the type $E = aW + b$, determine the values of a and b.
(c) What is the efficiency of this particular pulley block when raising its maximum permissible load of 2·5 kN?

***

Most relevant sections: 10.10, 10.2, and 10.5

***

C.G.L.I.

8. A screwjack has a single-start thread with a pitch of 8 mm. The effort is applied tangentially on a lever at a radius of 280 mm from the centre of the screw. Calculate:

(a) the velocity ratio of the screwjack,
(b) the load that can be lifted by an effort of 40 N if the efficiency is 25 per cent, and
(c) the work done by the operator during one revolution of the lever.

Giving a reason for your answer, state whether or not the screwjack would reverse, at this load, on removal of the effort.

***

Most relevant sections: 10.11, 10.5, 7.1, and 10.6

***

U.E.I.

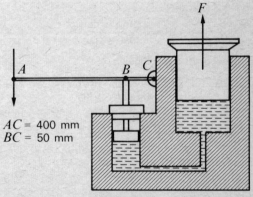

F

A B C

$AC = 400$ mm
$BC = 50$ mm

FIG. 10.21

9. (a) Distinguish between *force* and *pressure*.
(b) Fig. 10.21 shows part of a hydraulic jack. Neglecting losses, calculate the upward thrust F of the ram when a force of 30 N is applied at the end of

the operating handle *AC*. The ratio of the cross-sectional area of the ram to the cross-sectional area of the pump piston is 10:1.

Most relevant sections: 1.5 and 10.12

U.E.I.

10. Fig. 10.22 shows the elements of a small pelleting press for metal powders. Find the intensity of pressure, in MPa, acting on a pellet of diameter 14 mm when the force *F* is 400 N and both toggle arms lie at 10° to the vertical. Take π as $\frac{22}{7}$.
All frictional effects can be neglected.

Most relevant sections: 10.13 and 1.5

C.G.L.I.

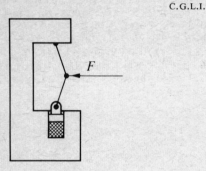

Fig. 10.22

11. A worm-and-wheel lifting device has a single-start worm driving a wheel with 50 teeth. The effort is applied through a pulley of diameter 120 mm and the load drum has a diameter of 200 mm. If an effort of 15 N is required to raise a load of 160 N, calculate:
(a) the velocity ratio, and
(b) the mechanical advantage and the efficiency at this particular loading.

Most relevant sections: 10.14, 10.3, and 10.5

U.E.I.

12. Fig. 10.23 shows the elements of a simple hand-operated arbor press. The rack has a pitch of 5 mm and the pinion has 33 teeth, the efficiency of the gearing being 75 per cent. An effort of 120 N is applied tangentially at a radius of 280 mm. Determine:
(a) the pitch circle diameter of the pinion, and
(b) the ram force.

Most relevant sections: 10.15 and 10.8.1

C.G.L.I.

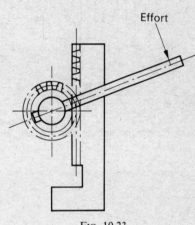

Fig. 10.23

13. (a) Determine the velocity ratio of the hand-operated geared winch shown diagrammatically in Fig. 10.24.
(b) The maximum load that should be lifted is 15 kN and it is found that at this loading the effort is 250 N. Calculate the mechanical advantage and the efficiency at this maximum loading.

Most relevant sections: 10.16, 10.3, and 10.5

C.G.L.I.

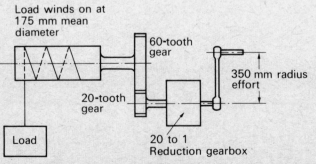

Fig. 10.24

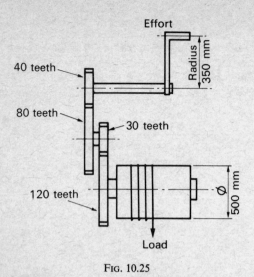

Effort

Radius 350 mm

40 teeth

80 teeth

30 teeth

120 teeth

∅ 500 mm

Load

FIG. 10.25

14. Fig. 10.25 shows the arrangement of a compound geared hand-winch. The radius of the effort handle

is 350 mm and the effective diameter of the load winding drum is 500 mm. Determine

(a) the velocity ratio of the winch,
(b) the mechanical advantage if a tangential effort of 250 N is required at the handle, to lift a load of 2 kN,
(c) the efficiency of the winch at this loading.

Most relevant sections: 10.8.2, 10.3, and 10.5

U.E.I.

ANSWERS TO EXERCISES 10

1. (a) 264, (b) 66, 25 per cent. **2.** (a) 1320, (b) 15·4 kN, (c) 58·3 N. **3.** (a) $a = 0·06$, $b = 5$, $E = 0·06\ W + 5$, (b) 1250 N, (c) 39·1 per cent. **4.** (a) 108 N m, (b) 3·394 kW, (c) 6·29 m/s. **5.** (a) 0·5 rev/s, (b) 1 m/s, (c) 4600 N m, (d) 14·4 kN. **6.** (a) 8, (b) 6·54, (c) 81·75 per cent. **7.** (a) 40, (b) $a = 0·05$, $b = 45$, (c) 36·76 per cent. **8.** (a) 220, (b) 2·2 kN, (c) 70·4 J. **9.** (b) 2·4 kN. **10.** 7·37 MPa. **11.** (a) 30, (b) 10·67, 35·6 per cent. **12.** (a) 52·51 mm, (b) 959·8 N. **13.** (a) 240, (b) 60, 25 per cent. **14.** (a) 11·2, (b) 8, (c) 71·3 per cent.

11 Heat and temperature

11.1. HEAT ENERGY

Energy, dealt with in Chapter 7, is defined as 'the capacity for doing work.' *Heat is one form of energy* and therefore provides this capacity for doing work. Heat can be defined as that kind of energy which is transferred across the boundary of a system because of a temperature difference between the system and the surroundings. The petrol engine and diesel engine are forms of 'heat engine' in which the appropriate fuel is burned to produce heat energy which is then converted into useful mechanical energy.

The following effects may be produced by the application of heat to a body:

(1) increase in temperature;
(2) increase in size;
(3) change in state;
(4) change in chemical constitution;
(5) change in electrical properties.

11.2. HEAT TRANSFER

Heat may be transferred from one place to another by three distinct processes. These are *conduction*, *convection*, and *radiation*.

11.2.1. Conduction. *Conduction* may be defined as *the transfer of heat through an unequally heated body from places of higher to places of lower temperature.*

This process of heat transfer continues until all parts of the body are at the same temperature.

This mode of heat transfer is usually concerned with solids. Conduction may be explained from energy considerations. Solids are made up of tightly packed molecules which are in continuous vibration about a fixed position. The application of heat at one place provides more energy to the molecules which then vibrate more violently. These greater vibrations are communicated to neighbouring molecules and so heat is transferred throughout the solid.

In general metals are good conductors, particularly silver, copper, and aluminium. Liquids are poor conductors and gases are very poor conductors of heat. Very poor conductors are termed insulators. Many porous substances, such as asbestos, granulated cork, glasswool, and firebrick are good insulators due to the inclusion of trapped air in the substance, air being a bad conductor.

Both insulators and conductors play useful parts in industrial processes. For example, in a steam-raising plant, to obtain both efficient and economical working the ratio of steam raised to fuel used must be as high as possible. To help to achieve this the following precautions are taken:

(1) The conduction of heat through the tubes is kept to a maximum by periodically removing the scale which forms on the tubes from dissolved solids in the water. This scale is a bad conductor.
(2) Steam pipes are lagged with an insulator such as asbestos to prevent heat loss.
(3) In the initial design, the grate or furnace in lined with insulating fire brick to reduce heat loss at this point.

11.2.2. Convection. *Convection* may be defined as *the transference of heat by the motion of the hot body itself carrying its heat with it.*

This mode of heat transfer takes place in liquids and gases.

When a liquid or gas is heated only part of the fluid near the heat source gets hot. This fluid expands, becomes less dense, and therefore rises. Cooler, and therefore denser, fluid takes its place and so a convection current is produced.

The cooling system of the motor car, and the heating and ventilation of buildings are direct applications of convection currents.

11.2.3. Radiation. *Radiation* may be defined as *the transfer of heat energy from a hot body to a cooler one without appreciable heating of the intervening space.*

This type of heat transfer is a wave motion similar to the transfer of light or radio waves. In fact radiated heat obeys all the laws of light waves.

Radiated heat unlike conducted and convected heat can pass through a vacuum. Heat radiated from the sun reaches the earth after travelling nearly 93 million miles through a vacuum.

In general when radiant heat falls on a material, part is absorbed, part reflected, and part transmitted. The ability of different materials and surfaces to absorb or radiate heat may be summarized as follows.

(1) The best absorbers and radiators of heat are dull black surfaces.
(2) Good absorbers are good radiators of heat and bad absorbers are bad radiators.
(3) Highly polished or silvered surfaces are good reflectors of heat but are poor absorbers.

11.3. TEMPERATURE

The degree of hotness of a body compared with some standard hotness is known as temperature. Temperature is independent of any substance.

It should be understood that temperature measures the

intensity of heat and not the quantity of heat contained by a body. Instruments used for temperature measurement are called thermometers.

11.4. TEMPERATURE SCALES

Although a variety of temperature scales have been employed in the past, only a few are now of importance. These are the celsius (formerly called centigrade), thermodynamic, and international temperature scales.

Note: The use of the name 'centigrade' for this purpose was abandoned by the Conference Generale des Poids et Mesures in 1948, and the British Standards Institution (B.S. 1991) recommends the use of 'degree Celsius'. In 1962 the celsius scale was adopted as the standard temperature scale in the U.K., but the thermodynamic scale (measurement in kelvins (K)) is now used in scientific work.

To establish a scale of temperature it is necessary to decide on fixed points which, for a particular substance have the same value at the same hotness.

11.4.1. Celsius temperature scale. *Anders Celsius* (1701–44) is credited with the introduction of the *celsius scale of temperature*. The fixed points on this scale, determined at a standard atmospheric pressure of 101 325 Pa, are:

Lower fixed point: freezing point (f.p.) of water: 0 °C
Higher fixed point: boiling point (b.p.) of water: 100 °C

To complete the scale of temperature the interval between 0 °C and 100 °C is divided into 100 divisions, each division representing 1 °C.

11.4.2. Thermodynamic temperature scale. The temperature scale which is independent of any thermometric substance is the *thermodynamic temperature scale*. This scale is due to *Lord Kelvin* (1824–1907) and has been adopted as the fundamental scale to which all temperature measurements should ultimately be referred. Being a theoretical scale, based on the laws of thermodynamics†, it is difficult to obtain in practice.

To place the thermodynamic temperature scale on a numerical basis, a fixed point called the triple point of water is given the value of 273·16 kelvin (K). The 'triple point (t.p.) of water' is the temperature of equilibrium between ice, water, and water vapour. The ice point (0 °C), or freezing point of water, is 273·15 K which for most practical purposes is taken as 273 K.

The number of divisions between the ice point and the boiling point (100 °C) on the kelvin scale is 100. This corresponds with the number of divisions between the same points on the Celsius scale.

Thus $T(\text{K}) = \theta\,(°\text{C}) + 273$.

A comparison of thermodynamic and celsius scales is given in Fig. 11.1 (also see Chapter 15).

† Thermodynamics is the name given to the science of the relationship between heat and mechanical energy.

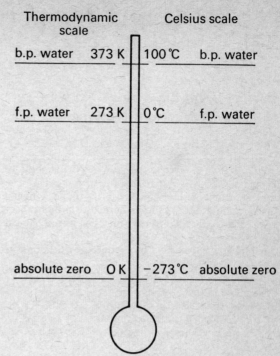

Thermodynamic scale		Celsius scale	
b.p. water	373 K	100 °C	b.p. water
f.p. water	273 K	0 °C	f.p. water
absolute zero	0 K	−273 °C	absolute zero

FIG. 11.1. Comparison of thermodynamic and celsius temperature scales

11.4.3. International practical temperature scale. The practical difficulties present in the measurement of temperature on the thermodynamic scale led to the introduction of a scale, easily and accurately reproducible, known as the international practical temperature scale (IPTS). This scale is expressed in degrees celsius and is, as far as possible, identical with the thermodynamic scale.

Thermometers for both laboratory and industrial purposes can be easily and accurately calibrated anywhere in the world by means of the IPTS. The scale is based on certain defining fixed points, all measured at standard atmosphere (101 325 Pa) by special thermometers. The 1968 revision reduced the lower limit of the scale to −259·34 °C, and some of the defining fixed points of the IPTS-68 are:

t.p. of equilibrium hydrogen	−259·34 °C
b.p. of neon	−246·048 °C
b.p. of oxygen	−182·962 °C
t.p. of water	0·01 °C
b.p. of water	100·00 °C
f.p. of zinc	419·58 °C
f.p. of silver	961·93 °C
f.p. of gold	1064·43 °C

11.4.4. Effect of pressure on the freezing point and boiling point of water. The fixed points for temperature scales must be determined at standard atmospheric pressure for the following reasons:

(1) an increase of pressure raises the boiling point and a decrease of pressure lowers the boiling point of water.

(2) an increase of pressure lowers the freezing point and a decrease of pressure raises the freezing point of water.

11.5. THERMOMETERS

Thermometers are devices which indicate temperature and depend upon one of the effects of heat, with the exception of 'change of state', for their successful operation.

The commonest type of thermometer is use is the mercury-in-glass thermometer. Since the freezing point of mercury is about $-39\,°C$, and the boiling point $+357\,°C$ at standard pressure, this thermometer is not used for temperatures much above $300\,°C$.

For temperatures below $-39\,°C$ thermometers containing alcohol may be used.

A suitable range of temperature for the alcohol-in-glass thermometer lies between $-130\,°C$ and $+60\,°C$.

Water is unsuitable as a thermometric liquid, not only because it freezes at $0\,°C$ but also on account of its irregular expansion.

In many industrial processes where high temperatures of $1000\,°C$ and above are frequently used instruments known as *pyrometers* are used for temperature measurement.

High temperatures may also be estimated with sufficient accuracy by reference to the known melting temperatures of certain substances. For example, the temperature of a body is approximately that of melting lead ($327\cdot3\,°C$) if a small piece of lead in contact with the body just melts. Substances used in this way are called *thermoscopes*.

11.5.1. Comparison of mercury and alcohol as thermometric substances.

Mercury

Advantages:

(1) it is opaque and can be clearly seen in a fine tube;
(2) its expansion is regular over a wide temperature range;
(3) it can easily be obtained in a pure state.

Disadvantages:

(1) its movement is sometimes rather jerky for it does not wet the glass tube.

Alcohol

Advantages:

(1) alcohol has a greater expansion than mercury for a given temperature rise; it is therefore more sensitive;
(2) it wets the tube and therefore does not stick to the glass.

Disadvantages:

(1) inaccuracies may occur due to the alcohol distilling over to the cooler part of the tube.

EXAMPLE 11.1. At standard atmospheric pressure the boiling point of oxygen is approximately $-183\,°C$. What is the corresponding temperature on the thermodynamic scale?

SOLUTION

$$T\,(\mathrm{K}) = \theta\,(°\mathrm{C}) + 273.$$

Then thermodynamic temperature $= (-183 + 273)\,\mathrm{K}$
$$= 90\,\mathrm{K}.$$

EXAMPLE 11.2. A quantity of air, originally at $15\,°C$, is heated so that its temperature is increased by $20\,°C$. What is its final temperature measured on the thermodynamic scale?

SOLUTION

Final temperature of air $= 15\,°C + 20\,°C$
$$= 35\,°C.$$
$$T\,(\mathrm{K}) = \theta\,(°\mathrm{C}) + 273.$$

Then thermodynamic temperature $= 35 + 273\,\mathrm{K}$
$$= 308\,\mathrm{K}.$$

EXAMPLE 11.3. At standard atmospheric pressure, the melting point of mercury is approximately $234\,\mathrm{K}$. Convert this temperature to $°C$.

SOLUTION

$$T\,(\mathrm{K}) = \theta\,(°\mathrm{C}) + 273.$$
$$\text{Then, } \theta = T - 273$$
$$= 234 - 273$$
$$= -39.$$

Temperature on celsius scale $= -39\,°C$.

EXAMPLE 11.4. Under standard atmospheric pressure, tungsten melts at $3380\,°C$ and boils at $5500\,°C$. What is the interval between these two temperatures expressed in (i) degrees Celsius, and (ii) kelvin?

SOLUTION

(i) Temperature interval $(\delta\theta) = (\theta_2 - \theta_1)\,°C$
$$= 5500 - 3380\,°C$$
$$= 2120\,°C.$$

(ii) In section 11.4.2 it has been shown that the number of divisions between corresponding points on the celsius and thermodynamic temperature scales is the same. It therefore follows that:

thermodynamic temperature interval $(\delta T) = 2120\,\mathrm{K}$.

SUMMARY

Heat is one form of energy and can be defined as that kind of energy which is transferred across the boundary of a system because of a temperature difference between the system and the surroundings.

Conduction is the term which describes heat transfer through solids. Heat energy is passed from molecule to molecule throughout the solid mass. In general, metals are good conductors. Many porous substances are good insulators due to the inclusion of trapped air which is a bad conductor.

Convection may be defined as the transfer of heat by the motion of the hot body itself carrying its heat with it. This method of heat transfer takes place in liquids and gases.

Radiation may be defined as the transfer of heat energy from a hot body to a cooler one without appreciable heating of the intervening space. Radiated heat will pass through a vacuum.

Temperature is defined as the degree of hotness of a body compared with some standard hotness. Temperature is the *intensity* of heat.

$$T(\text{K}) = \theta\,(^\circ\text{C}) + 273.$$

EXERCISES 11

1. Explain, briefly, the difference between heat and temperature. State two properties of matter which would enable it to be used for measuring temperatures.

Most relevant sections: 11.1 and **11.3**

2. What are the advantages and disadvantages of alcohol as compared with mercury for use in a thermometer? Give reasons why water is unsuitable as a thermometric liquid.

Most relevant sections: 11.5 and **11.5.1**

3. Explain the following terms: thermometer, scale of temperature, fixed points.

Most relevant sections: 11.4 and **11.5**

4. A length of copper wire is heated from 20 °C to its melting point of 1083 °C. List the changes that take place in the wire during the process.

Most relevant section: 11.1

5. (a) Name the three methods of transferring heat from one body to another.
   (b) Why is very little heat transferred by any of these methods in a thermos flask?

Most relevant section: 11.2

C.G.L.I.

6. Distinguish between the transmission of heat by conduction, convection, and radiation, giving one example of each. What precautions are taken to reduce heat losses from steam raising plant?

Most relevant section: 11.2

7. Listed below are materials which may either be used as a good conductor or a good insulator. Tabulate these materials in their respective groups.
   Silver, aluminium, glass, tin, wood, cork, iron, air, asbestos, and copper.

Most relevant section: 11.2

8. State the effects of changes in pressure on the boiling point and freezing point of water.

Most relevant section: 11.4.4

9. A quantity of air, originally at 20 °C, is heated so that its temperature increases by 40 °C. What is its final temperature measured on the thermodynamic scale?

Most relevant section: 11.4.2

10. The melting point of nitrogen, at standard atmospheric pressure, is −210 °C. What is the equivalent temperature on the thermodynamic scale?

Most relevant section: 11.4.2

11. At standard atmospheric pressure, the boiling point of neon is approximately 27 K. Convert this temperature to degrees Celsius.

Most relevant section: 11.4.2

12. Under standard atmospheric pressure, titanium melts at 1680 °C and boils at 3300 °C. What is the interval between these two temperatures expressed in (a) degrees Celsius, and (b) kelvin?

Most relevant section: 11.4.2

ANSWERS TO EXERCISES 11

9. 333 K. **10.** 63 K. **11.** −246 °C. **12.** (a) 1620 °C, (b) 1620 K.

12 Quantity of heat

12.1. QUANTITY OF HEAT

In order to produce a change of temperature in a body, heat energy must either be supplied to it or removed from it.

When a temperature change takes place in a body the *quantity of heat* lost or gained depends upon three factors:

(a) the *mass* of the body,
(b) the *temperature change* of the body, and
(c) the *specific heat capacity* of the material of the body.

The relationship between these quantities is:

Quantity of heat
(lost or gained) =

$$\frac{\text{mass}}{\text{of body}} \times \frac{\text{specific heat capacity}}{\text{of body}} \times \text{temperature change.}$$

Using symbols,

$$\delta Q = m \times c \times \delta\theta,$$

where m = mass of the body, c = specific heat capacity, and $\delta\theta$ = temperature change.

In SI units heat, like all other forms of energy, is measured in joules (J).

12.2. HEAT CAPACITY

Two kilograms of a given substance require twice as much heat as one kilogram of that substance to raise their temperatures by the same amount.

Heat (or thermal) capacity is defined as *the quantity of heat required to raise the temperature of a body by one degree*. The heat capacity of a substance will depend upon the mass and nature of the substance.

Heat capacity of a body = mass × specific heat capacity.

The unit of heat capacity is the joule per kelvin (J/K).

12.3. SPECIFIC HEAT CAPACITY

Different solids, liquids, and gases require different amounts of heat to produce the same temperature rise. For instance, water requires more heat to produce a certain temperature rise than an equal mass of any other solid or liquid. The heat capacity of unit mass of a substance is called the specific heat capacity.

The specific heat capacity of a substance is defined as the heat required to raise the temperature of unit mass of the substance by one degree.

It will be seen from the relationship $\delta Q = m \times c \times \delta\theta$ that in SI units,

$$\text{specific heat capacity, } c = \frac{\delta Q \text{ (J)}}{m \text{(kg)} \times \delta\theta \text{ (°C or K)}}.$$

Thus the unit of specific heat capacity is the joule per kilogram kelvin (J/kg K).

TABLE 12.1

Approximate values for the specific heat capacity of some common substances

Substance	Specific heat capacity (kJ/kg K)	Substance	Specific heat capacity (kJ/kg K)
Alcohol, ethyl	2·29	Mercury	0·138
Aluminium	0·88	Nickel	0·461
Brass	0·377	Paraffin	2·18
Brine	2·89	Platinum	0·134
Copper	0·398	Solder	0·176
Glass	0·796	Steel	0·481
Glycerine	2·43	Tin	0·231
Gold	0·13	Turpentine	1·76
Iron (cast)	0·503	Water	4·187
Lead	0·126	Zinc	0·386

Bearing in mind the definition of specific heat capacity consider the following: From Table 12.1 the specific heat capacity of aluminium equals 0·88 kJ/kg K and of turpentine equals 1·76 kJ/kg K. Then,

(1) 0·88 kJ of heat are needed to raise the temperature of 1 kg of aluminium through 1 K,
(2) 1·76 kJ of heat are needed to raise the temperature of 1 kg of turpentine through 1 K.

It will be seen, for equal masses of aluminium and turpentine, that the amount of heat required to raise the temperature of turpentine through a given temperature interval is twice the amount of heat needed to raise the temperature of aluminium through the same temperature interval. It therefore follows that if equal masses of both these substances were supplied with heat at the same rate, the time required to raise the temperature of the turpentine through a certain interval would be twice the time needed to raise the temperature of the aluminium through the same temperature interval.

EXAMPLE 12.1. Calculate the amount of heat required to raise the temperature of: (a) 20 l of water from 20 °C to boiling point (100 °C) and (b) 30 kg of ethyl alcohol from 20 °C to boiling point (78·4 °C).
Specific heat capacity of water = 4·2 kJ/kg K.
Specific heat capacity of ethyl alcohol = 2·3 kJ/kg K.

SOLUTION

(a) Quantity of heat
$$\begin{array}{c}\text{(kJ)}\end{array} = \frac{\text{mass}}{\text{(kg)}} \times \frac{\text{specific heat capacity}}{\text{(kJ/kg K)}} \times \frac{\text{temperature rise}}{\text{(°C or K)}}.$$

Let quantity of heat required = δQ.

Assume 120 l of water = 120 kg,

then $\delta Q = 120 \times 4\cdot2 \times (100-20)$.

$\therefore \delta Q = 40\,320$ kJ

$= 40\cdot32$ MJ.

Amount of heat required to raise temperature of 120 l of water from 20 °C to 100 °C is 40·32 MJ.

(b) Quantity of heat
$$\begin{array}{c}\text{(kJ)}\end{array} = \frac{\text{mass}}{\text{(kg)}} \times \frac{\text{specific heat capacity}}{\text{(kJ/kg K)}} \times \frac{\text{temperature rise}}{\text{(°C or K)}}.$$

Let quantity of heat required = δQ,

then $\delta Q = 30 \times 2\cdot3 \times (78\cdot4-20)$

$= 4029\cdot6$ kJ

$= 4\cdot0296$ MJ.

Amount of heat required to raise temperature of 30 kg of ethyl alcohol from 20 °C to 78·4 °C is 4·0296 MJ.

EXAMPLE 12.2. A small furnace produces 15 MJ of heat per hour. How long will it take to heat 30 kg of copper from 20° C to 180 °C? (Take specific heat capacity of copper = 0·4 kJ/kg K.)

SOLUTION

The first step is to calculate the total amount of heat required to heat the 30 kg of copper from 20 °C to 180 °C.

Using $\delta Q = mc\,\delta\theta$

$= 30 \times 0\cdot4 \times (180-20)$

$= 1920$ kJ

$= 1\cdot92$ MJ.

Now the furnace produces 15 MJ per hour:

therefore time taken for 1·92 MJ to be produced equals

$$\frac{1\cdot92}{15} \times 60 \text{ min}$$

$= 7\cdot7$ min.

EXAMPLE 12.3. 20 kg of water at 100 °C has the temperature reduced to 65 °C. How much heat energy is lost by the water during this process? Take the specific heat capacity of water as 4·2 kJ/kg K.

If 20 kg of fluid of specific heat capacity 2·5 kJ/kg K at 65 °C had received the same amount of heat energy as that lost by the water, what would have been the final temperature of the fluid?

U.L.C.I.

SOLUTION

Part 1

Let the heat energy lost = δQ

Quantity of heat
$$\begin{array}{c}\text{lost (kJ)}\end{array} = \frac{\text{mass}}{\text{(kg)}} \times \frac{\text{specific heat capacity}}{\text{(kJ/kg K)}} \times \frac{\text{fall in temperature}}{\text{(°C or K)}}$$

$\delta Q = 20 \times 4\cdot2 \times (100-65)$

$= 2940$ kJ

$= 2\cdot94$ MJ.

Heat lost = 2·94 MJ.

Part 2

Let final temperature of the fluid = θ.

From Part 1, heat given to the fluid = 2·94 MJ or 2940 kJ.

Temperature rise of the fluid = $(\theta-65)$ °C.

Using $\delta Q = mc\,\delta\theta$,

$2940 = 20 \times 2\cdot5 \times (\theta-65)$,

$\dfrac{2940}{20 \times 2\cdot5} = \theta-65$,

or $\theta - 65 = \dfrac{2940}{50}$,

$\theta = 65 + \dfrac{2940}{50}$,

$\therefore \qquad \theta = 65 + 58\cdot8 = 123\cdot8$.

Final temperature of fluid = 123·8 °C.

EXAMPLE 12.4. How much heat is needed to change the temperature of a steel tank which contains 800 kg of water from 12 °C to 92 °C? The tank has a mass of 100 kg, the specific heat capacity of the steel is 0·5 kJ/kg K, and the specific heat capacity of water is 4·2 kJ/kg K.

SOLUTION

Let total amount of heat required (i.e. for tank and water) = δQ.

Heat absorbed by tank
$$= \frac{\text{mass of}}{\text{tank (kg)}} \times \frac{\text{specific heat capacity}}{\text{of steel (kJ/kg K)}} \times \frac{\text{temperature rise}}{\text{(°C or K)}}$$

$= 100 \times 0\cdot5 \times (92-12)$

$= 4000$ kJ

$= 4$ MJ.

Heat absorbed by water
$$= \frac{\text{mass of}}{\text{water (kg)}} \times \frac{\text{specific heat capacity}}{\text{of water (kJ/kg K)}} \times \frac{\text{temperature rise}}{\text{(°C or K)}}$$

$= 800 \times 4\cdot2 \times 80$ kJ

$= 268\,800$ kJ

$= 268\cdot8$ MJ.

Then $\delta Q = 4 + 268\cdot8$ MJ

$= 272\cdot8$ MJ.

Quantity of heat required = 272·8 MJ.

12.4. CALORIFIC VALUE OF FUEL

Every fuel whether solid, liquid, or gaseous has a heat energy content which is released when the fuel is burned. The calorific value of a fuel is defined as *the number of heat units released when unit mass or volume of the fuel is burned.* For instance, the calorific value of petrol is approximately 37·6 MJ/l. This means that 37·6 MJ of heat energy are released when 1 litre of petrol is burned.

The approximate calorific values of some of the more common fuels are given in the following table.

TABLE 12.2

Fuel	Calorific value	
	Mass basis	*Volume basis*
Anthracite	36·4 MJ/kg	—
Bituminous coal	33·5 MJ/kg	—
Diesel oil	46·2 MJ/kg	36·8 MJ/l
Gas, blast-furnace	—	3·6 MJ/m³
Gas, natural	—	34·1 MJ/m³
Gas, town	—	19·0 MJ/m³
Paraffin	46·8 MJ/kg	37·2 MJ/l
Petrol	47·1 MJ/kg	37·6 MJ/l

12.5. MIXTURE OF HOT AND COLD BODIES

Previous examples in this chapter have been concerned with either a heat loss or a heat gain.

The following problems are concerned with the transfer of heat between two bodies in contact when they are at different temperatures.

If it is assumed that the heat lost to the surroundings is negligible then the following equation may be used.

Heat lost by hot body = **heat gained** by cold body,

where **heat lost** = mass × specific heat capacity × **fall** in temperature by hot body,

and **heat gained** = mass × specific heat capacity × **rise** in temperature by cooler body

EXAMPLE 12.5. A block of steel of mass 50 kg is heated to 765 °C and then quenched in a tank containing 800 l of oil at 15 °C. What is the final temperature of the oil? (Neglect all heat losses to the tank and surroundings).

Specific heat capacity of steel = 0·5 kJ/kg K.

Specific heat capacity of oil = 2 kJ/kg K.

Relative density of oil = 0·8.

SOLUTION

Let θ be the final temperature of the oil.

The final temperature of the steel will also be θ. Then,

fall in temperature of the steel = $(765-\theta)$ °C,
rise in temperature of the oil = $(\theta-15)$ °C.

Heat **lost** by steel = mass × specific heat capacity × **fall** in temperature

$$= 50 \times 0·5 \times (765-\theta) \text{ kJ}$$
$$= 25 (765-\theta) \text{ kJ}$$

Heat **lost** by steel = $(19\,125 - 25\,\theta)$ kJ.

Heat **gained** by oil = mass × specific heat capacity × **rise** in temperature

$$= (800 \times 0·8) \times 2 \times (\theta-15) \text{ kJ}$$

(*Note:* Mass of 1 litre of water = 1 kg
∴ Mass of 1 litre of oil = 0·8 kg.)

$$= 640 \times 2 \times (\theta-15) \text{ kJ}$$
$$= 1280 (\theta-15) \text{ kJ}$$

Heat **gained** by oil = $(1280\,\theta - 19\,200)$ kJ.

Using

heat **lost** by steel = heat **gained** by oil
$$19\,125 - 25\,\theta = 1280\,\theta - 19\,200$$
$$38\,325 = 1305\,\theta$$
$$\theta = \frac{38\,325}{1305}$$
$$= 29·3 \text{ °C.}$$

Final temperature of the oil = 29·3 °C.

EXAMPLE 12.6. In an experiment to determine the specific heat capacity of iron the following results were obtained:

Mass of calorimeter (copper)	= 90 g.
Combined mass of calorimeter and water	= 230 g.
Combined mass of calorimeter, water, and iron	= 420 g.
Initial temperature of water	= 17 °C.
Initial temperature of iron	= 97 °C.
Final temperature of iron and water	= 27 °C.
Specific heat capacity of copper	= 0·4 kJ/kg K.
Specific heat capacity of water	= 4 kJ/kg K.

Calculate the specific heat capacity of the iron.

SOLUTION

Let the specific heat capacity of iron equal c.

Fall in temperature of the iron = $97-27 = 70$ °C.

Rise in temperature of water and calorimeter = $27-17 = 10$ °C.

Mass of water = $230-90$ = 140 g or 0·14 kg.

Mass of iron = $420-230$ = 190 g or 0·19 kg.

Heat lost by iron = mass (kg) × specific heat capacity of iron (kJ/kg K) × fall in temperature (°C or K)

$$= 0·19 \times c \times 70 \text{ kJ}$$
$$= 13·3\,c \text{ kJ}$$

Similarly,

heat gained by calorimeter $= 0.09 \times 0.4 \times 10$ kJ
$$= 0.36 \text{ kJ},$$

and, heat gained by water $= 0.14 \times 4 \times 10$ kJ
$$= 5.6 \text{ kJ}.$$

Using, heat lost = heat gained (neglecting heat losses),
$$13.3 \, c = 0.36 + 5.6$$
$$c = \frac{5.96}{13.3}$$
$$= 0.45.$$

Specific heat capacity of iron $= 0.45$ kJ/kg K.

12.6. THERMODYNAMICS

Thermodynamics is the name given to *the science of the relationship between heat and mechanical energy*. The principle of the conservation of energy (Chapter 7) states that 'energy can neither be created nor destroyed but may be converted from one form into another'. In accordance with this principle there would thus appear to be some connexion between heat and mechanical energy, and the first law of thermodynamics is concerned with this connexion.

12.6.1. First law of thermodynamics. The *law of equivalence of heat and work* may be stated as follows:

Heat and mechanical work are mutually convertible, and in any action involving such conversion a definite quantity of heat disappears for every unit of work done, and, conversely a definite quantity of mechanical energy is used for every unit of heat generated.

The relationship between work and heat is known as Joule's equivalent of heat (J). Whenever W units of mechanical work are completely used in generating Q units of heat, $J = W/Q$. In the SI, because the unit of heat is also the unit of work, the value of J is 1 Nm/J, and its inclusion in equations is unnecessary and even redundant.

12.6.2. Efficiency of conversion. In all operations involving the conversion of heat energy into mechanical work there is a large waste: not all the energy available in the fuel is usefully converted into the required form. The thermal efficiency of a heat engine may be of the order of 15 per cent to 30 per cent since it is extremely difficult to prevent heat being 'lost'. Some of the heat produced by the combustion of the fuel will be wasted in radiation, cooling water, exhaust gases, windage and friction, etc.

The process of converting mechanical and other forms of energy into heat is not accompanied by such an excessive waste; e.g. an electric heater may have an efficiency of about 80 per cent to 90 per cent.

EXAMPLE 12.7. A steel drill driven at 3.5 rev/s makes a hole in a mild steel block and causes a temperature rise of 25 °C in $\frac{1}{2}$ min. If the block has a mass of 5 kg and the drill has a mass of 0.5 kg, calculate the torque on the drill and the power expended. The specific heat capacity of steel is 500 J/kg K.

SOLUTION

$$\begin{aligned}\text{Heat absorbed by} \atop \text{drill and block} &= {\text{total mass} \atop \text{of steel}} \times {\text{specific heat} \atop \text{capacity}} \times {\text{tempera-} \atop \text{ture rise}}\\ &= 5.5 \times 500 \times 25 \text{ J}\\ &= 68\,750 \text{ J}.\end{aligned}$$

$$\begin{aligned}{\text{Heat generated} \atop \text{per second}} &= \frac{68\,750}{30} \text{ J/s}\\ &= 2292 \text{ J/s}\end{aligned} \qquad (1)$$

But 1 J/s = 1 W,

∴ power expended $= 2.292$ kW.

If T N m be the torque and n rev/s the speed of rotation, the work done

$$\begin{aligned}\text{per second} &= 2\pi nT\\ &= 2\pi \times 3.5 \times T = 22T \text{ J}.\end{aligned} \qquad (2)$$

Equating (1) and (2):

$$2292 = 22T.$$
$$\therefore \qquad T = \frac{2292}{22}$$
$$= 104.2,$$
$$\therefore \qquad \text{torque} = 104.2 \text{ N m}.$$

EXAMPLE 12.8. A certain engine develops 25 kW for a fuel consumption of 9 kg of oil per hour. If the calorific value of the oil is 47 MJ/kg, what percentage of the heat supplied in the oil is converted into useful work?

SOLUTION

$$\begin{aligned}\text{Heat supplied per hour} &= 9 \times 47 \text{ kJ}\\ &= 423 \text{ MJ}.\end{aligned}$$

$$\begin{aligned}\text{Heat supplied per second} &= \frac{423\,000}{3600} \text{ kJ}\\ &= 117.5 \text{ kJ}.\end{aligned}$$

$$\text{Power developed} = 25 \text{ kW}$$

i.e. work done per second = 25 kJ/s.

$$\begin{aligned}\text{Thermal efficiency} &= \frac{\text{work done per second}}{\text{energy supplied per second}}\\ &= \frac{25}{117.5}\\ &= 0.213.\end{aligned}$$

Expressed as a percentage the thermal efficiency is 21.3 per cent.

EXAMPLE 12.9. A shaft carrying a dead load of 3 t is supported in plain cylindrical bearings of diameter 150 mm. If the shaft revolves at 20 rev/s, and the coefficient of friction between the shaft and bearings is 0.03 calculate:

(a) the frictional torque,
(b) the power absorbed in friction, and
(c) the heat generated per minute by the friction.
  Take g as 9.81 m/s².

SOLUTION

(*Note*: The frictional resistance R is tangential to the shaft and opposes the motion as shown by Fig. 12.1.)

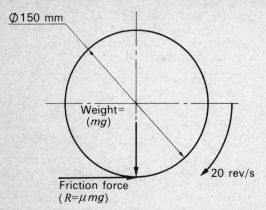

Fig. 12.1

(a)
$$R = \mu mg$$
$$= 0{\cdot}03 \times 3000 \times 9{\cdot}81$$
$$= 882{\cdot}9.$$

Friction force $= 882{\cdot}9$ N.

Diameter of shaft $= 150$ mm $= 0{\cdot}15$ m, so that

frictional torque $(= Rr) = 882{\cdot}9 \times \dfrac{0{\cdot}15}{2}$ N m

$$= 66{\cdot}3 \text{ N m.}$$

(b) Shaft revolves at 20 rev/s.

Power $= 2\pi n T$ (where n is in rev/s) J/s

$$= 2\pi \times 20 \times 66{\cdot}3 \text{ J/s}$$
$$= 8370 \text{ J/s or W.}$$

(c) Heat generated/min $=$ work done against friction/min

$$= 8370 \times 60 \text{ J}$$
$$= 502\,200 \text{ J}$$
$$= 502{\cdot}2 \text{ kJ.}$$

SUMMARY

Quantity of heat (lost or gained) $=$ mass $\times$ specific heat capacity $\times$ temperature change

or

$$\delta Q = mc\delta\theta.$$

The specific heat capacity of a substance is defined as the heat required to raise the temperature of unit mass of the substance by 1 degree (°C or K).

Specific heat capacity $(c) = \dfrac{\delta Q \text{ (J)}}{m \text{ (kg)} \times \delta\theta \text{ (°C or K)}}$ J/kg K.

Heat capacity of a body $=$ mass $\times$ specific heat capacity. The unit of heat capacity is the joule per kelvin (J/K).

Calorific value of a fuel is defined as the number of heat units released when unit mass or volume of the fuel is burned.

Method of mixtures:

$\dfrac{\text{Heat lost by hot body}}{\text{(neglecting losses).}} =$ Heat gained by cold body.

A definite quantity of heat can be produced by the expenditure of a given amount of mechanical energy.

The relation between work and heat is known as Joule's mechanical equivalent of heat (J).

In the SI, J is 1 N m/J and its inclusion in equations is unnecessary.

EXERCISES 12

1. 4 kg of water at 89·5 °C is mixed with 3 kg of water at 30 °C. What will be the final temperature of the mixture?

Most relevant section: 12.5

W.J.E.C.

2. State, using the appropriate units in each case, the amount of heat required in each of the following cases:

(a) To raise the temperature of 50 kg of water of specific heat capacity 4·2 kJ/kg K by 40 °C ....
(b) To raise the temperature of 5 kg of oil of specific heat capacity 2·1 kJ/kg K from 10 °C to 50 °C ....

Most relevant section: 12.1

U.E.I.

3. A piece of aluminium alloy of mass 2·5 kg has a specific heat capactiy of 1 kJ/kg K. After being heated to a high temperature it was quickly immersed in some water contained in a calorimeter. The water had a mass of 6 kg and the water equivalent of the calorimeter was 1·5 kg. Assuming that no heat was lost, calculate the temperature of the aluminium alloy before immersion if the initial temperature of the water and calorimeter was 15 °C and the final temperature after immersion became 38 °C. Specific heat capacity of water is 4·2 kJ/kg K.

Most relevant section: 12.5

U.E.I.

4. (a) Briefly describe an experiment you have carried out to determine the specific heat of a substance.
(b) In such an experiment the following results were recorded:

Mass of calorimeter	= 65 g.
Combined mass of calorimeter and water	= 305 g.
Combined mass of calorimeter, water and specimen	= 425 g.
Initial temperature of water	= 20 °C.
Initial temperature of specimen	= 95 °C.
Final temperature of water and specimen	= 25 °C.
Specific heat capacity of calorimeter metal	= 0·42 kJ/kg K.
Specific heat capacity of water	= 4·2 kJ/kg K.

Calculate the specific heat capacity of the specimen.

Most relevant section: 12.5

U.E.I.

5. Define (a) the joule, (b) the specific heat capacity of a substance. How much heat is lost by 55 kg of water cooling from 100 °C to 40 °C? Take the specific heat capacity of water as 4·2 kJ/kg K.

Most relevant sections: 12.1 and 12.3

N.C.T.E.C.

6. A piece of steel of mass 4 kg has a specific heat capacity of 0·5 kJ/kg K. It is heated to a high temperature and at this temperature it is immersed in some water in a calorimeter. The initial temperature of the water and calorimeter was 20 °C. The final temperature, after immersion, became 60 °C. The water in the calorimeter had a mass of 5 kg and the water equivalent of the calorimeter was 1·5 kg. Assuming that there were no heat energy losses, calculate the temperature of the steel before immersion.
Specific heat capacity of water is 4·2 kJ/kg K.

Most relevant section: 12.5

U.L.C.I.

7. 20 kg of water at 100 °C has the temperature reduced to 20 °C. How much heat energy is lost by the water during this process? Specific heat capacity of water is 4·2 kJ/kg K.
If 20 kg of fluid of specific heat capacity 2·5 kJ/kg K at 20 °C had received the same amount of heat energy as that lost by the water, what would have been the final temperature of the fluid?

Most relevant section: 12.1

U.L.C.I.

8. A piece of metal of mass 4 kg and specific heat capacity 0·63 kJ/kg K, has its temperature raised from 100 °C to 800 °C. How much heat transfer is required to carry out this temperature change?
What is the water equivalent of the piece of metal?

Most relevant section: 12.1

U.L.C.I.

9. (a) State *three* ways in which the addition of heat can affect a substance.
(b) A piece of metal of mass 12 kg absorbs 45 kJ when its temperature is raised from 15 °C to 100 °C. Find the specific heat capacity of the metal.

Most relevant sections: 11.1 and 12.1

N.C.T.E.C.

10. A zinc casting of mass 50 kg is placed in a bath containing 180 kg of brine. As a result of the immersion, the temperature of the casting is raised by 75 °C. The bath is made of steel and is of mass 120 kg. Calculate (a) the quantity of heat transferred to the zinc; (b) the total water equivalent of the brine and bath; and (c) the temperature drop of the brine and the bath.
Specific heat capacities are: Zinc = 0·4 kJ/kg K; brine = 2·9 kJ/kg K; steel = 0·5 kJ/kg K; water = 4·2 kJ/kg K.

Most relevant sections: 12.1 and 12.5

N.C.T.E.C.

11. In an experiment, a piece of copper of mass 24·5 g was immersed in boiling water at 100 °C and then transferred to a calorimeter containing 121 g water

at 12·0 °C. The final temperature of the water was 13·6 °C. Calculate the value for the specific heat capacity of copper given by these results.

Specific heat capacity of water is 4·2 kJ/kg K.

What are the likely sources of error in this experiment?

Most relevant section: 12.5

C.G.L.I.

12. A piece of steel of mass 60 kg is to be cooled from 500 ° to 100 °C by lowering it into a tank containing oil. If the average temperature rise of the oil is to be limited to 65 °C what is the smallest mass of oil which could be used? Neglect losses and take the specific heat capacities of steel and oil as 0·5 and 1·9 kJ/kg K respectively.

Most relevant section: 12.5

Y.C.F.E.

13. The frictional loss in an engine bearing is known to be 750 W. Calculate the heat generated by friction per hour.

Most relevant sections: 7.3 and 12.6.1

E.M.E.U.

14. A piece of lead falls 3 m from rest, coming to rest again on the ground. Calculate the rise in its temperature. Take the specific heat capacity of lead as 134 J/kg K and g as 9·81 m/s².

Most relevant sections: 9.4.1, 12.1, and 12.6.1

15. Find the mass of petrol used per hour by an engine working at 4 kW, if the calorific value of the fuel is 45 MJ/kg and if 20 per cent of this heat is converted into useful work.

Most relevant sections: 12.4 and 12.6.1

16. A shaft carrying a dead load of 90 kg, is supported in plain cylindrical bearings of 60 mm diameter.

If the coefficient of friction between the shaft and bearings is 0·02, and the shaft revolves at 25 rev/s calculate:

(a) the tangential force due to friction;
(b) the friction torque;
(c) the power absorbed in friction;
(d) the heat generated per minute by the friction in the bearings.

Most relevant sections: 6.4, 9.8, and 12.6.1

U.E.I.

17. The input to a gearbox is supplied as a tangential load of 1·2 kN on a spur gear of effective diameter 450 mm which runs at 18 rev/s.

If the output shaft runs at 6 rev/s and the transmission efficiency is 95 per cent, what will be the torque and power on the output shaft?

Also, if all the power lost in the gear box is converted into heat, express this loss in kJ/min.

Most relevant sections: 7.3, 9.7, and 12.6.1

U.E.I.

18. Explain briefly the term 'mechanical equivalent of heat'. An engine uses 10 kg of oil per hour and has a thermal efficiency of 25 per cent. Find the power developed if the heating value of the oil is 46 MJ/kg. Calculate the mass of cooling water required per minute if 50 per cent of the heat from the fuel enters the cooling water and the increase in temperature is not to exceed 45 °C. Take specific heat capacity of water as 4·2 kJ/kg K.

Most relevant sections: 12.6.1, 7.3, and 12.1

E.M.E.U.

19. A 375 W electric motor has a steel drum fitted to its shaft, and a friction brake presses against this drum. The mass of the drum is 3 kg and the specific heat capacity of its material 420 J/kg K. If the initial temperature of the drum is 15 °C, find its temperature after the motor has been running for 1 min, assuming that all the work done by the motor is converted into heat and that there is no loss of heat from the drum.

Most relevant sections: 7.3, 12.1, and 12.6.1

20. (a) In a heat-treatment operation a steel component, of mass 5 kg and specific heat capacity 480 J/kg K, is heated to a temperature of 830 °C and then quenched in oil. The oil has a specific heat capacity of 2400 J/kg K and an initial temperature of 20 °C. If the temperature of the oil must not exceed 30 °C, and heat losses can be neglected, calculate the minimum mass of oil that should be used.

(b) If 25 per cent of the heat absorbed by the oil can be recovered and components are quenched once every 5 min, calculate the power which can be recovered.

Most relevant sections: 12.5 and 7.3

U.E.I.

21. (a) A rectangular block of steel measures 200 mm × 80 mm × 50 mm. If the density of the steel is 7800 kg/m³, calculate its mass.

(b) During a hardening process the steel block was heated from 20 °C to 840 °C. In order to bring about the necessary temperature rise the furnace used 0·15 m³ of natural gas with a calorific value of 34 MJ/

m³. If the specific heat capacity of the steel is 480 J/kg K, calculate the thermal efficiency of the furnace.

Most relevant sections: 1.3.1, 12.1, and 12.4

U.E.I.

ANSWERS TO EXERCISES 12

1. 64 °C. **2. (a)** 8·4 MJ, **(b)** 420 kJ. **3.** 328 °C. **4.** 0·62 kJ/kg K. **5.** 13·9 MJ. **6.** 606 °C. **7.** 6·72 MJ, 154 °C. **8.** 1·76 MJ, 0·6 kg. **9.** 44 J/kg K. **10.** 1·5 MJ, 138 kg, 2·6 °C. **11.** 0·382 kJ/kg K. **12.** 97 kg. **13.** 2·7 MJ. **14.** 0·22 °C. **15.** 1·6 kg. **16. (a)** 17·7 N, **(b)** 0·53 N m, **(c)** 84 W, **(d)** 5 kJ. **17.** 770 N m, 29·1 kW, 90 kJ/min. **18.** 32 kW, 20·3 kg/min. **19.** 32·9 °C. **20. (a)** 80 kg, **(b)** 1·6 kW. **21. (a)** 6·24 kg, **(b)** 48·2 per cent.

13 Heat and change of state

13.1. SENSIBLE HEAT

Sensible heat is that heat which *causes a change of temperature in a substance*. The term 'sensible' used in this way means 'able to be observed'. For example, a thermometer records a change of temperature when sensible heat is given to or given up by a body.

EXAMPLE 13.1. Calculate the sensible heat given to a mass of 10 kg of water in raising its temperature from 10 °C to 50 °C. The specific heat capacity of water is taken as 4·2 kJ/kg K.

SOLUTION

Heat required to raise the temperature of water from 10 °C to 50 °C

$$= \text{mass} \times \frac{\text{specific heat}}{\text{capacity}} \times \frac{\text{temperature rise from}}{\text{10 °C to 50 °C}}$$

$$= 10 \text{ kg} \times 4·2 \frac{\text{kJ}}{\text{kg K}} \times 40 \text{ K} \qquad \begin{array}{l}\text{(Temperature interval of} \\ \text{40 °C} = \text{numerically} \\ \text{temperature interval of} \\ \text{40 K)}\end{array}$$

$$= 1680 \text{ kJ}.$$

∴ Sensible heat given to 10 kg of water in raising its temperature by 40 °C = 1·68 MJ.

13.2. CHANGE OF STATE

All substances at ordinary temperatures and pressures exist in one of three forms or states, namely, solid, liquid, or gaseous. However, many substances undergo a change of state when sufficient heat energy is either supplied to them or removed from them. For example, water at normal temperatures exists in the liquid state, but removal of sufficient heat energy produces the solid state: ice. Similarly, if sufficient heat energy is given to water a change from the liquid to the gaseous state (steam) occurs. Certain substances, however, require both a change of temperature and pressure before a change of state takes place.

13.2.1. Melting or fusing point. When a solid reaches a certain temperature, further heating will not increase the temperature of the substance until it has changed completely into a liquid state. This particular temperature is called the *melting point* or *fusing point* of the substance.

13.3. ENTHALPY OF FUSION (LATENT HEAT OF FUSION)

13.3.1. Change of enthalpy. If a substance undergoes a reversible process at constant pressure, e.g. if it melts, the quantity of heat required to bring about the change of state is equal to the change of *enthalpy* (see 13.3.2 below).

Enthalpy is a concept used in the study of heat, and its origin is beyond the scope of this book.

The SI unit of enthalpy is that of energy, i.e. the joule (J).

13.3.2. Solid state to liquid. When a solid has been raised to its melting point it can only be melted by supplying a definite amount of heat to it. This heat is absorbed by the substance without any change of temperature occurring; it is used entirely for changing the state of the substance from solid to liquid.

The heat absorbed by the solid in this manner is equal to the change of enthalpy, and is called the 'enthalpy of fusion' or the 'latent heat of fusion'. The word 'latent' means 'hidden' and indicates that although heat is absorbed no temperature change takes place.

Note: The latent heat energy given to a substance is used entirely in overcoming the strong forces of attraction which exist between the molecules of the solid.

The specific enthalpy of fusion or specific latent heat of fusion of a substance is defined as the amount of heat required to convert unit mass of a substance at its melting point into liquid at the same temperature.

The symbol used to denote specific enthalpy of fusion is h_{if}. The subscript *if* means *solid to liquid* (fusion). The letter *i* in the subscript signifies 'saturated solid', while the letter *f* indicates 'saturated liquid'.

13.3.3. Units. Specific enthalpy or specific latent heat is expressed in heat units per unit mass; the preferred unit being kJ/kg.

For example, the specific enthalpy of fusion of ice (see Table 13.1) may be expressed as 335 kJ/kg.

The quantity of heat required to change a solid at its melting point, into liquid at the same temperature = mass of solid × specific enthalpy (specific latent heat) of fusion.

Similarly, the quantity of heat given out when a liquid solidifies without change of temperature = mass of liquid × specific enthalpy (specific latent heat) of fusion.

13.3.4. Liquid to solid state. This is the reverse process to that previously described. When a liquid reaches a certain temperature, further cooling will not reduce the temperature of the substance until it has completely changed into a solid state. This particular temperature is called the *freezing point* of the substance. It is equal in value to the melting point temperature.

The heat energy given up by a substance at its freezing

point, without change in temperature, is equal to the enthalpy (latent heat) of fusion of the substance.

Consider the changes of state from ice to water and water to ice. These changes may be shown as follows:

Solid to liquid:

$$\begin{array}{c}\text{1 kg ice}\\\text{at 0 °C}\end{array} + \begin{array}{c}\text{335 kJ}\\\text{(latent heat)}\end{array} \rightarrow \begin{array}{c}\text{1 kg water}\\\text{at 0 °C.}\end{array}$$

Liquid to solid:

$$\begin{array}{c}\text{1 kg water}\\\text{at 0 °C}\end{array} - \begin{array}{c}\text{335 kJ}\\\text{(latent heat)}\end{array} \rightarrow \begin{array}{c}\text{1 kg ice}\\\text{at 0 °C.}\end{array}$$

The following table gives the *specific enthalpy (specific latent heat)* of fusion of some common substances.

TABLE 13.1

Substance	Specific enthalpy of fusion or specific latent heat of fusion h_{lf} (kJ/kg)
Aluminium	403
Cadmium	54·5
Copper	205
Ice (fresh water)	335
Iron	268
Lead	25·2
Mercury	11·7
Tin	58·8
Zinc	101

EXAMPLE 13.2. Calculate the amount of heat required to melt 50 kg of zinc at its melting point temperature. (Specific enthalpy, or specific latent heat, of fusion of zinc is 101 kJ/kg.)

SOLUTION

Quantity of heat required to melt 1 kg of zinc at its melting point = 101 kJ.
Quantity of heat required to melt 50 kg of zinc at its melting point

$$= 50 \times 101 \text{ kJ}$$
$$= 5050 \text{ kJ}$$
$$= 5·05 \text{ MJ.}$$

13.4. ENTHALPY OF EVAPORATION (LATENT HEAT OF VAPORIZATION)

13.4.1. Liquid state to vapour. It has already been stated that to change a substance, at its melting point, from the solid to the liquid state requires a certain quantity of latent heat energy.

In a similar way the change of state from liquid to vapour takes place with the absorption of heat by the liquid without a temperature change taking place. The

heat absorbed by the liquid in this manner is equal to the change of enthalpy, and is called the 'enthalpy of evaporation' or 'latent heat of vaporization' (or evaporation).

The specific enthalpy of evaporation, or specific latent heat of vaporization (or evaporation), is defined as the quantity of heat required to change unit mass of a substance from the liquid state into vapour without change of temperature.

The symbol used to denote the specific enthalpy of evaporation is h_{fg}. The subscript *fg* means *liquid to vapour* (evaporation) where *f* again represents 'saturated liquid' and *g* stands for 'dry saturated vapour'.

The quantity of heat required to change liquid into vapour at the same temperature

= mass of liquid × specific enthalpy (specific latent heat) of evaporation.

Similarly, the quantity of heat given out when vapour condenses without change of temperature

= mass of vapour × specific enthalpy (specific latent heat) of evaporation.

13.4.2. Vapour to liquid state. The process of changing from vapour to liquid is known as *condensation*.

When a substance changes from vapour to the liquid state all the latent heat of vaporization is given up before the vapour is completely converted into liquid at the same temperature.

The following table gives the specific enthalpy (specific latent heat) of vaporization or evaporation of a number of substances.

TABLE 13.2

Substance	Specific enthalpy of evaporation or specific latent heat of vaporization, h_{fg} (kJ/kg)
Alcohol, ethyl	855
Butane	386
Carbon tetrachloride	193
Helium	25·2
Hydrogen	453
Mercury	273
Oxygen	214
Sulphuric acid	511
Water (steam)	2260

EXAMPLE 13.3. How much heat could be obtained from 10 kg of dry saturated steam at atmospheric pressure if it were converted to water at 60 °C? (Specific enthalpy of

evaporation for steam, or specific latent heat of steam, is 2·26 MJ/kg. Specific heat capacity of water is 4·2 kJ/kg K.)

SOLUTION

Enthalpy given up by 1 kg of steam at 100 °C on conversion to 1 kg water at 100 °C = 2·26 MJ.

∴Enthalpy given up by 10 kg of steam
$$= 2·26 \times 10 = 22·6 \text{ MJ.}$$

Sensible heat given up by water on cooling from 100 °C to 60 °C = mass × fall of temperature
$$\times \text{ specific heat capacity}$$
$$= 10 \times (100 - 60) \times 4·2 \text{ kJ}$$
$$= 1680 \text{ kJ}$$
$$= 1·68 \text{ MJ.}$$

Total heat given out by

steam and water = enthalpy + sensible heat
$$= 22·6 + 1·68 \text{ MJ}$$
$$= 24·28 \text{ MJ.}$$

13.5. EVAPORATION AND BOILING

Evaporation and boiling are both concerned with the change of state from liquid to vapour.

Evaporation has the following characteristics:

(1) It takes place at all temperatures from the surface of a liquid.
(2) The rate of evaporation may be increased by either increasing the surface area or increasing the temperature of the liquid. A draught of air across the surface of a liquid also accelerates evaporation.
(3) Evaporation has the effect of cooling a liquid since the enthalpy of evaporation is supplied from the body of the liquid.
(4) The lower the temperature of a liquid the greater is the total heat required for complete evaporation.

The cooling effect produced by the evaporation of a liquid mentioned in (3) above is made use of in the design of commercial refrigerators.

Boiling takes place in the body of a liquid and occurs at one temperature, namely, the *boiling point*. In this case the enthalpy of evaporation is supplied by some external source.

EXAMPLE 13.4. Calculate the amount of heat energy required to convert 100 g of ice at −5 °C into steam at 100 °C.

Specific enthalpy (specific latent heat) of fusion of ice = 335 kJ/kg.
Specific enthalpy (specific latent heat) of evaporation of water = 2260 kJ/kg.
Specific heat capacity of ice   = 2·1 kJ/kg K.
Specific heat capacity of water = 4·2 kJ/kg K.

SOLUTION

The total heat required to convert ice at −5 °C into steam at 100 °C will be dealt with in four stages as follows:

1. Sensible heat required to raise ice at −5 °C to ice at 0 °C equals
   mass of ice × specific heat capacity × rise of temperature
   $= 0·1 \times 2·1 \times 5$ kJ (Temperature rise from −5 °C to 0 °C = 5 °C.)
   $= 1·05$ kJ.

2. Enthalpy (latent heat) required to change ice at 0 °C into water at 0 °C equals
   mass of ice × specific enthalpy of fusion of ice
   $= 0·1 \times 335$ kJ
   $= 33·5$ kJ.

3. Sensible heat required to raise water at 0 °C to water at 100 °C equals
   mass of water × specific heat capacity × rise of temperature
   $= 0·1 \times 4·2 \times (100 - 0)$ kJ
   $= 42$ kJ.

4. Enthalpy (latent heat) required to change water at 100 °C into steam at 100 °C equals
   mass of water × specific enthalpy of evaporation of water
   $= 0·1 \times 2260$ kJ
   $= 226$ kJ.

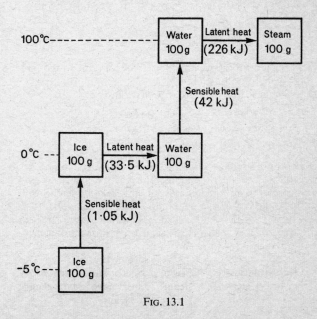

FIG. 13.1

Total heat required to convert 100 g of ice at $-5\,°C$ into steam at $100\,°C$ equals

$$1\cdot05 + 33\cdot5 + 42 + 226$$
$$= 302\cdot55\text{ kJ.}$$

Fig. 13.1 illustrates the stages of conversion from ice at $-5\,°C$ into steam at $100\,°C$.

13.6. DRYNESS FRACTION OF STEAM

The change of state of a fluid from the liquid state to dry saturated vapour takes place with the absorption of heat by the fluid at constant temperature. At normal atmospheric pressure water boils at $100\,°C$ and steam is produced. At any other given value of the pressure there is a corresponding boiling point. Until the full amount of enthalpy has been supplied to the water to cause complete vaporization, the steam generated remains in contact with the water and minute particles of water will be held suspended in the steam. Such steam is called *wet steam*. When the water has been completely converted into steam, the resulting vapour is called *dry saturated steam*. Wet steam is visible due to the presence of the water particles (it is the water particles which are seen). Dry steam is invisible since the change of state is complete.

The *dryness fraction x* of steam (or the quality of steam) is the ratio:

$$\frac{\textbf{mass of dry steam}}{\textbf{total mass of wet steam containing it}}.$$

Consider Fig. 13.2 which shows a mass $(m_s + m_w)$ of wet steam in which, for convenience, all the water particles are assumed to be at the bottom of the container. The water particles have a mass m_w so that the dry steam has a mass m_s. The dryness fraction of the steam is, therefore, given by

$$x = \frac{m_s}{m_s + m_w}.$$

In general, the above equation holds for a mixture of any dry vapour and its liquid.

The heat in a quantity of wet steam is given by:

$$\frac{\textbf{heat in}}{\textbf{wet steam}} = \frac{\textbf{sensible}}{\textbf{heat}} + \frac{\textbf{dryness}}{\textbf{fraction}} \times \textbf{enthalpy}$$

13.7. SUPERHEATED STEAM

If heat is supplied to dry saturated steam (or to any dry saturated vapour) the temperature of the dry steam will increase because the change of state is complete. Steam that is heated to a temperature above the temperature at which it is generated is called *superheated steam*. The heat supplied is calculated in the same way as for sensible heat, that is:

$$\frac{\textbf{heat transferred}}{\textbf{to dry steam}} = \frac{\textbf{mass of}}{\textbf{steam}} \times \frac{\substack{\textbf{specific heat}\\\textbf{capacity of}\\\textbf{superheated}\\\textbf{steam}}}{} \times \frac{\textbf{tempera-}}{\textbf{ture change}}$$

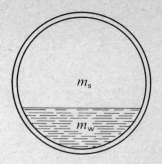

Fig. 13.2

The total heat in a quantity of superheated steam is illustrated by Fig. 13.3 and is given by:

$$\frac{\textbf{total heat in}}{\textbf{superheated steam}} = \frac{\textbf{sensible}}{\textbf{heat}} + \textbf{enthalpy} + \textbf{superheat}$$

In practice steam is either wet or superheated. A given mass of superheated steam contains more energy than the same mass of wet steam, and as a result superheated steam has many important uses such as in turbine plants.

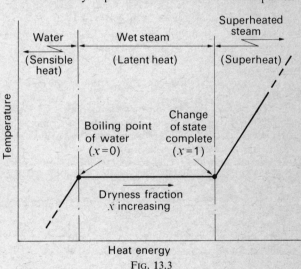

Fig. 13.3

EXAMPLE 13.5. Assuming atmospheric pressure, find the amount of heat energy required to convert 5 kg of feed water at $20\,°C$ into:

(a) 5 kg of wet steam, dryness fraction 0·9, and
(b) 5 kg of superheated steam at $300\,°C$.

Specific heat capacity of water $= 4\cdot2\text{ kJ/kg K.}$

Specific heat capacity of superheated steam $= 2\cdot1\text{ kJ/kg K.}$

Specific enthalpy (latent heat) of evaporation of water $= 2260\text{ kJ/kg.}$

SOLUTION

(a) Heat transferred = sensible heat + $\dfrac{\text{dryness}}{\text{fraction}}$ × enthalpy

$$= 5\,\text{kg} \times 4\cdot2\,\frac{\text{kJ}}{\text{kg K}} \times (100 - 20)\,°\text{C}$$

$$+ 0\cdot9 \times 5\,\text{kg} \times 2260\,\frac{\text{kJ}}{\text{kg}}$$

$$= 1680\,\text{kJ} + 10\,170\,\text{kJ}$$

$$= 11\,850\,\text{kJ}$$

$$= 11\cdot85\,\text{MJ}.$$

(b) Heat transferred = sensible heat + enthalpy + superheat

$$= 1680\,\text{kJ} + 5\,\text{kg} \times 2260\,\frac{\text{kJ}}{\text{kg}}$$

$$+ 5\,\text{kg} \times 2\cdot1\,\frac{\text{kJ}}{\text{kg K}} \times (300 - 100)\,°\text{C}$$

$$= 1680\,\text{kJ} + 11\,300\,\text{kJ} + 2100\,\text{kJ}$$

$$= 15\,080\,\text{kJ}$$

$$= 15\cdot08\,\text{MJ}.$$

SUMMARY

Sensible heat causes a change in temperature in a body.

Enthalpy (latent heat) causes a change in state in a body at constant temperature.

(1) The specific enthalpy of fusion, or specific latent heat of fusion, of a substance is the quantity of heat which must be given at steady temperature to unit mass of the substance in order to change its state from solid to liquid.

(2) The specific enthalpy of evaporation, or specific latent heat of vaporization, of a substance is the quantity of heat which must be given at steady temperature to unit mass of the substance in order to change its state from liquid to vapour.

$$\frac{\text{Sensible}}{\text{heat}} = \text{mass} \times \frac{\text{specific heat}}{\text{capacity}} \times \frac{\text{change in}}{\text{temperature}}.$$

$$\frac{\text{Enthalpy}}{\text{(latent heat)}} = \frac{\text{mass} \times \text{specific enthalpy}}{\text{(specific latent heat)}}.$$

$$\frac{\text{Dryness}}{\text{fraction}}{\text{of steam}} = \frac{\text{mass of dry steam}}{\text{total mass of wet steam containing it}}.$$

$$\frac{\text{Heat in}}{\text{wet steam}} = \frac{\text{sensible}}{\text{heat}} + \frac{\text{dryness}}{\text{fraction}} \times \text{enthalpy}.$$

$$\frac{\text{Heat transferred}}{\text{to dry steam}} = \frac{\text{mass of}}{\text{steam}} \times \frac{\text{specific heat}}{\text{capacity of}}{\text{superheated}}{\text{steam}} \times \frac{\text{temperature}}{\text{change}}.$$

$$\frac{\text{Total heat in}}{\text{superheated}}{\text{steam}} = \frac{\text{sensible}}{\text{heat}} + \text{enthalpy} + \text{superheat}.$$

EXERCISES 13

1. Determine the amount of heat to be removed from 1 kg of water at 16 °C to produce ice at a temperature of −4 °C.
   Specific enthalpy (latent heat) of fusion of ice = 335 kJ/kg.
   Specific heat capacity of ice = 2·1 kJ/kg K.
   Specific heat capacity of water = 4·2 kJ/kg K.

Most relevant sections: **13.1** and **13.3.3**

N.C.T.E.C.

2. What is meant by the statement 'specific enthalpy of steam at atmospheric pressure is 2·26 MJ/kg'? Convert this figure to kJ/kg.

Most relevant section: **13.4**

C.G.L.I.

3. When 20 g of ice at 0 °C is thoroughly mixed in a calorimeter containing water the final temperature is 10 °C. The total water equivalent of the calorimeter and water before the ice is added is 100 g. Determine the initial temperature of the water. The specific enthalpy (latent heat) of fusion of ice is 335 kJ/kg. Assume no losses. Specific heat capacity of water is 4·2 kJ/kg K.

Most relevant sections: **13.3.3** and **12.5**

Y.C.F.E.

4. The specific enthalpy (latent heat) of steam at a pressure of 0·1 MPa is 2·26 MJ/kg. Calculate the total heat required to convert 30 kg of water at a temperature of 18 °C and pressure 0·1 MPa into dry steam. Specific heat capacity of water is 4·2 kJ/kg K.

Most relevant sections: **13.1** and **13.4**

U.E.I.

5. How much heat could be obtained from 5 kg of dry saturated steam at atmospheric pressure if it were converted to water at 60 °C?
   (Specific enthalpy (latent heat) of steam = 2·26 MJ/kg at atmospheric pressure. Specific heat capacity of water = 4·2 kJ/kg K.)

Most relevant sections: **13.1** and **13.4**

U.E.I.

6. Calculate the heat required to raise a lump of lead having a mass of 5 kg from 15 °C to melting point and then just to melt it without further increase of temperature.
(Specific heat capacity = 0·13 kJ/kg K; specific enthalpy (latent heat) of fusion = 25 kJ/kg; melting point = 327 °C.)
If this heat were put into 25 kg of water at 12 °C to what value would its temperature rise? Assume that no heat is lost. Specific heat capacity of water = 4·2 kJ/kg K.

Most relevant sections: 13.1 and 13.3

U.L.C.I.

7. (a) What is meant by:
(i) sensible heat,
(ii) latent heat (enthalpy).

(b) 3 kg of water is contained in a vessel having a water equivalent of 0·15 kg, and is at a temperature of 40 °C. 0·1 kg of dry saturated steam is injected into the water. Assuming no heat to be lost, what will be the final temperature of the water?
Take specific enthalpy (latent heat) of steam = 2260 kJ/kg. Specific heat capacity of water = 4·2 kJ/kg K.

Most relevant sections: 13.1, 13.3, 13.4, and 12.5

U.E.I.

8. A piece of ice of mass 30 g at a temperature of −8 °C is immersed in a copper calorimeter of mass 100 g and containing 400 g of water at 25 °C. Find the final temperature of the water assuming no heat losses.

Specific heat capacity of copper =
0·4 kJ/kg K.
Specific heat capacity of ice =
2·1 kJ/kg K.
Specific enthalpy (latent heat) of fusion of ice =
335 kJ/kg.
Specific heat capacity of water =
4·2 kJ/kg K.

Most relevant sections: 13.3 and 12.5

E.M.E.U.

9. Define enthalpy (latent heat) and sensible heat.
Find the quantity of steam at atmospheric pressure that is required to change 200 g of ice at −6 °C into water at 50 °C. The specific enthalpy (latent heat) of vaporization of steam is 2·26 MJ/kg and the specific latent heat of fusion of ice is 335 kJ/kg. The specific heat capacity of ice is 2·1 kJ/kg K and the specific heat capacity of water is 4·2 kJ/kg K.

Most relevant sections: 13.1, 13.3, 13.4, and 12.5

E.M.E.U.

10. (a) A block of aluminium alloy is in the form of a rectangular prism 200 mm × 80 mm × 50 mm. If the density of the alloy is 2800 kg/m³, calculate its mass in kilograms.
(b) Taking the specific heat capacity of the alloy to be 1000 J/kg °C, how much energy, in kilojoules, is needed to raise the temperature of the block from 160 °C to its melting point of 660 °C?
(c) Taking the specific latent heat of the alloy as 400 kJ/kg, how much further energy is required, in kilojoules, to melt the block whilst its temperature remains at 660 °C?

Most relevant sections: 1.3.1, 12.1, and 13.3

C.G.L.I.

11. (a) What is meant by (i) sensible heat and (ii) latent heat (enthalpy)?
(b) How much heat energy must be transferred from 10 kg of dry saturated steam, at atmospheric pressure, if it is to be converted into water at 60 °C? Take the specific latent heat of steam (or specific enthalpy of evaporation for steam) as 2300 kJ/kg and the specific heat capacity of water as 4·2 kJ/kg K.

Most relevant sections: 12.1 and 13.4

U.E.I.

12. (a) Distinguish between the *evaporation* and *boiling* of a liquid.
(b) Calculate the amount of heat required to raise the temperature of 5 kg of copper from 23 °C to its melting point of 1083 °C and then to melt it without further increase of temperature. Take the specific heat capacity of copper as 0·4 kJ/kg K and the specific (latent heat) of fusion (or specific enthalpy of fusion) of copper as 205 kJ/kg.

Most relevant sections: 12.5, 12.1, and 13.3

U.E.I.

13. An aluminium alloy has a specific heat capacity of 900 J/kg K, a melting point of 650 °C and a specific latent heat of fusion of 400 × 10³ J/kg.

(a) Calculate the heat energy required to completely melt a 2 kg mass of this alloy if its initial temperature is 20 °C.

(b) An electric furnace is to be used to melt the alloy and the complete operation is to be completed in 12 min. Determine the power rating of the furnace, in kilowatts, assuming that only 70 per cent of the available power is usefully used to melt the alloy.

Most relevant sections: 12.1, 13.3, and 7.4

U.E.I.

14. (a) A factory uses waste steam to heat a liquid used for a plating process. The steam enters a heat exchanger at atmospheric pressure, its dryness fraction being 0·9. The steam leaves the heat exchanger converted to water at 40 °C. How much heat energy is available for the process per kilogram of wet steam? Take the specific latent heat of vaporization at atmospheric pressure to be 2260 kJ/kg, and the specific heat capacity of water to be 4·2 kJ/kg K.

(b) The process consists of heating a quantity of liquid having a specific heat capacity of 3·77 kJ/kg K and a mass of 230 kg. The container for the liquid has a mass of 90 kg and has a specific heat capacity of 0·5 kJ/kg K. The temperature of the liquid has to rise from 12 °C to 24 °C. Neglecting heat losses how many kilograms of wet steam are required for the process?

Most relevant sections: 13.6 and 12.1

C.G.L.I.

15. (a) Tin has a melting point of 230 °C, a specific heat capacity of 0·23 kJ/kg K, and a specific latent heat of 58·7 kJ/kg. What quantity of heat is required to melt 22·7 kg of tin originally at 30 °C?

(b) Steel has a specific heat capacity of 0·46 kJ/kg K. What quantity of heat is necessary to raise 1·6 kg of steel from 30 °C to 230 °C?

(c) If the steel object referred to in (b) at its original temperature of 30 °C is plunged into the molten tin referred to in (a), what mass of tin solidifies when the steel object and the tin both acquire a temperature of 230 °C?

Most relevant sections: 12.1, 13.3, and 12.5

C.G.L.I

16. Calculate the heat energy required to produce 10 kg of superheated steam at 250 °C and atmospheric pressure, from feed water at 60 °C.
Specific heat capacity of water = 4·2 kJ/kg K.
Specific heat capacity of superheated steam = 2·1 kJ/kg K.
Specific enthalpy (latent heat) of vaporization of steam = 2260 kJ/kg.

Most relevant section: 13.7

ANSWERS TO EXERCISES 13

1. 410·6 kJ. **2.** 2260 kJ/kg. **3.** 28 °C. **4.** 78·1 MJ. **5.** 12·1 MJ. **6.** 328 kJ; 15·1 °C. **7.** 60 °C. **8.** 18·2 °C. **9.** 45 g. **10.** (a) 2·24 kg, (b) 1120 kJ, (c) 896 kJ. **11.** 24 680 kJ. **12.** (b) 3145 kJ. **13.** (a) 1934 kJ, (b) 3·83 kW. **14.** (a) 2286 kJ, (b) 4·79 kg. **15.** (a) 2375 kJ, (b) 147 kJ, (c) 2·51 kg. **16.** 27·43 MJ.

14 Heat and expansion

14.1. EXPANSION

Most solids, liquids, and gases expand when the temperature is increased, and contract when cooled. During the expansion of a given body its mass remains constant and its volume increases. Hence its density, i.e. mass per unit volume, decreases.

In engineering, expansion has many advantages and disadvantages. The steel tyre for a locomotive wheel is bored slightly smaller than the diameter of the wheel; expansion takes place when the tyre is heated and it may be fitted on to the wheel. The contraction of the cooling tyre binds the assembly together. A steam pipe becomes longer when steam passes through it, and some method of preventing the pipe from buckling must be employed. An expansion bend (Fig. 14.1) may be included in the pipeline to allow movement to take place without damage occurring.

The amount of expansion in a substance depends upon three factors:

(1) *the original size;*
(2) *the temperature rise;*
(3) *the nature of the substance.*

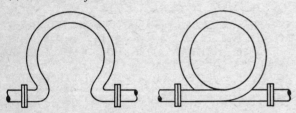

FIG. 14.1. Expansion bends

14.2. LINEAR EXPANSION

It is often only necessary to take account of expansion in one direction, i.e. length. The amount of expansion which occurs in a given direction is proportional to the original size, so that a body expands by a maximum amount along its length (Fig. 14.2).

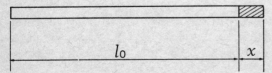

FIG. 14.2. Linear expansion

14.2.1. Thermal coefficient of linear expansion. *The thermal coefficient of linear expansion of a substance may be defined as the change in length which unit length of the substance undergoes when its temperature is changed by 1 degree.*

From an inspection of Table 14.1 it can be seen that the value of the thermal coefficient of linear expansion for aluminium is $0 \cdot 000\ 025\ 5/°C$. This figure means that:

1 m of aluminium expands by $0 \cdot 000\ 025\ 5$ m per degree temperature rise;

1 km of aluminium expands by $0 \cdot 000\ 025\ 5$ km per degree temperature rise; etc.

Let l_0 = the original length of the material,

α = the thermal coefficient of linear expansion of the material, and

$\delta\theta$ = the change of temperature.

Then, increase per unit length = $\alpha\ \delta\theta$,

total increase $x = l_0 \alpha\ \delta\theta$,

and final length $l = l_0 + l_0 \alpha\ \delta\theta$,

$= l_0 (1 + \alpha\ \delta\theta)$.

It can be seen that for a fall in temperature the change in length must be subtracted from the original length.

TABLE 14.1

Average values of thermal coefficients of linear expansion

Material	Coefficient α (per °C)
Aluminium	$25 \cdot 5 \times 10^{-6}$
Brass	$18 \cdot 9 \times 10^{-6}$
Copper	$16 \cdot 7 \times 10^{-6}$
Iron (cast)	$10 \cdot 2 \times 10^{-6}$
Iron (wrought)	$11 \cdot 9 \times 10^{-6}$
Lead	$29 \cdot 1 \times 10^{-6}$
Nickel	$12 \cdot 8 \times 10^{-6}$
Steel (mild)	$11 \cdot 9 \times 10^{-6}$
Tin	$21 \cdot 4 \times 10^{-6}$
Zinc	$26 \cdot 3 \times 10^{-6}$

14.2.2. Bimetallic devices. A bimetallic strip consists of a compound strip of two metals, mechanically joined face-to-face by rivetting, welding, or other means. Fig. 14.3(a) shows a bimetallic strip of two metals P and Q, P having a higher thermal coefficient of linear expansion than Q.

When the temperature rises, the mean length of P becomes greater than that of Q, so that the strip bends as shown by Fig. 14.3(b). This variation of curvature due to change in temperature is employed in appliances such as thermostats for the control of ovens and furnaces, fire alarms, etc.

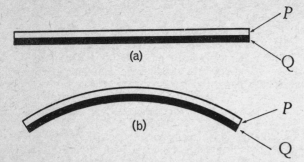

FIG. 14.3. Bimetallic strip

EXAMPLE 14.1. The thermostat switch for an electrical immersion heater is operated by a brass rod. The rod is 300 mm long at 12 °C and the switch opens when the water temperature reaches 85 °C. Given that the thermal coefficient of linear expansion for the brass is $18·9 \times 10^{-6}$ per °C, calculate the length of the rod when the switch opens.

SOLUTION

The length of the rod at 85 °C is given by
$$l = l_0(1 + \alpha \delta\theta),$$
where $\delta\theta = 85\ °C - 12\ °C = 73\ °C.$
Then $l = 300\ (1 + 18·9 \times 10^{-6} \times 73)$
$$= 300\ (1 + 0·001\ 38)$$
$$= 300·414.$$
The length $= 300·414$ mm.

EXAMPLE 14.2. A mild steel sphere of diameter 50 mm rests on an aluminium ring having an internal diameter of 49·9 mm. What minimum increase in temperature is required for the sphere to pass through the ring? Take the thermal coefficients of linear expansion for steel and aluminium as $11·9 \times 10^{-6}$ per °C and $25·5 \times 10^{-6}$ per °C respectively.

SOLUTION

(*Note:* Fig. 14.4 shows the arrangement. For the sphere to pass just through the ring the diameters must be equal. An increase in temperature is common to both the sphere and ring. Diameter is a linear dimension.)

For the sphere:
$$\text{new diameter } d = d_0(1 + \alpha \delta\theta)$$
$$= 50\ (1 + 11·9 \times 10^{-6}\ \delta\theta).$$
For the ring:
$$\text{new diameter } d = 49·9\ (1 + 25·5 \times 10^{-6}\ \delta\theta).$$

But the change in temperature $\delta\theta$ is common.

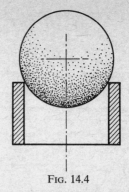

FIG. 14.4

Then
$$50(1 + 11·9 \times 10^{-6}\ \delta\theta) = 49·9(1 + 25·5 \times 10^{-6}\ \delta\theta),$$
$$50 + 595 \times 10^{-6}\ \delta\theta = 49·9 + 1272·45 \times 10^{-6}\ \delta\theta,$$
$$50 - 49·9 = 10^{-6}\ \delta\theta\ (1272·45 - 595),$$
$$0·1 = 677·45 \times 10^{-6}\ \delta\theta,$$
∴ minimum increase in temperature $= 147·6$ °C.

EXAMPLE 14.3. A steel collar is to be shrunk on to a shaft 67·5 mm diameter, to fit tightly when the collar is 84 °C above the temperature of the shaft. Calculate the diameter to which the collar must be bored. In order that the collar may be fitted it is expanded to give an all round clearance of 0·15 mm. Estimate the temperature change through which the collar must be heated. The thermal coefficient of linear expansion for the steel is $11·9 \times 10^{-6}$ per °C.

SOLUTION

Note: The collar is to have an internal diameter of 67·5 mm when its temperature is 84 °C above its initial temperature. Diameter is a linear dimension.)
From $d = d_0(1 + \alpha\delta\theta),$

$$d_0 = \frac{d}{1 + \alpha\delta\theta}$$
$$= \frac{67·5}{1 + (11·9 \times 10^{-6} \times 84)}$$
$$= \frac{67·5}{1·001}$$
$$= 67·44$$
∴ original bore of collar $= 67·44$ mm.

An all round clearance of 0·15 mm gives a total of 0·3 mm on the diameter.

Then, required collar expansion $(x) = 67·8 - 67·44$ mm
$$= 0·36\ \text{mm}.$$
But $x = d_0\alpha\delta\theta,$
$$0·36 = 67·44 \times 11·9 \times 10^{-6} \times \delta\theta,$$
∴ temperature change $\delta\theta = 448$ °C.

14.3. SUPERFICIAL EXPANSION

In dealing with expansion in two directions, i.e. length and breadth, it can be seen that a change in area takes place (Fig. 14.5).

Superficial expansion is considered when dealing with metal plates and sheets, etc.

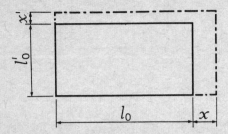

FIG. 14.5. Area expansion

14.3.1. Thermal coefficient of superficial expansion. *The thermal coefficient of superficial expansion of a substance may be defined as the change in area which unit area of the substance undergoes when its temperature is changed by one degree.*

The thermal coefficient of superficial expansion for a given material is twice the coefficient of linear expansion for the material. If follows that the change in area of a lamina of the material is given by $2\alpha A_0\delta\theta$, where A_0 is the original area, from which:

$$\text{the final area } A = A_0 + 2\alpha A_0\delta\theta$$
$$= A_0(1 + 2\alpha\delta\theta).$$

14.4. CUBICAL EXPANSION

When the temperature of a body is raised, expansion takes place in all directions so that an increase in volume is experienced (Fig. 14.6).

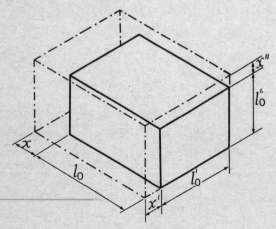

FIG. 14.6. Volume expansion

14.4.1. Thermal coefficient of cubical expansion. *The thermal coefficient of cubical expansion of a substance may be defined as the change in volume which unit volume of the substance undergoes when its temperature is changed by one degree.*

The thermal coefficient of cubical expansion for a given material is three times the coefficient of linear expansion for the material. Hence the change in volume of a given body is $3\alpha V_0\delta\theta$ where V_0 is the original volume, so that the final volume V

$$= V_0 + 3\alpha V_0\delta\theta,$$
$$= V_0(1 + 3\alpha\delta\theta).$$

EXAMPLE 14.4. A rectangular block of zinc measures 20 mm × 5 mm × 4 mm. Find the change in volume of the block if its temperature is raised by 200 °C. The thermal coefficient of linear expansion of zinc is $26\cdot3 \times 10^{-6}$ per °C.

SOLUTION

$$\text{Original volume of block } (V_0) = 20 \times 5 \times 4 \text{ mm}^3$$
$$= 400 \text{ mm}^3.$$

Thermal coefficient of cubical

$$\text{expansion } (\beta) = 3\alpha$$
$$= 3 \times 26\cdot3 \times 10^{-6} \text{ per}°\text{C}.$$
$$\text{Increase in volume} = V_0\beta\delta\theta$$
$$= 400 \times 78\cdot9 \times 10^{-6} \times 200 \text{ mm}^3$$
$$= 6\cdot312 \text{ mm}^3.$$

SUMMARY

The expansion of a body depends upon:

(1) the original size;
(2) the temperature rise; and
(3) the nature of the substance.

Linear expansion is expansion considered in one direction. The length may be large compared with the width and thickness, e.g. railway lines and overhead electrical cables.

Thermal coefficient of linear expansion is the increase in length per unit length per degree temperature rise.

$$l = l_0(1 + \alpha\delta\theta).$$

Superficial expansion is expansion considered in two directions. The thickness may be small compared with the length and width, e.g. plates and sheets of material.

Thermal coefficient of superficial expansion is the increase in area per unit area per degree temperature rise.

$$A = A_0(1 + 2\alpha\delta\theta).$$

Cubical expansion is expansion considered in all directions. No dimension may be small enough to be neglected, e.g. blocks of material.

Thermal coefficient of cubical expansion is the increase in volume per unit volume per degree temperature rise.

$$V = V_0(1 + 3\alpha\delta\theta).$$

EXERCISES 14

1. A steel gauge has a length of 150 mm at 20 °C. Calculate the length at a temperature of 100 °C. The thermal coefficient of linear expansion of steel = 0·000 011 per °C.

Most relevant section: 14.2.1

E.M.E.U.

2. A bar of metal 10 m long at 15 °C increases in length by 15 mm when heated to a temperature of 100 °C. Calculate the 'thermal coefficient of linear expansion' for the metal. State the units of this 'coefficient'.

Most relevant section: 14.2.1

E.M.E.U.

3. A steel shaft is exactly 0·5 m diameter at 12 °C. What is the new diameter at 32 °C? Thermal coefficient of linear expansion of steel is 0·000 011 per °C. Give one example of the expansion of a metal being (a) an advantage (b) a disadvantage.

Most relevant sections: 14.1 and 14.2.1

C.G.L.I.

4. (a) A collar has an outside diameter of 8 cm, an inside diameter of 6 cm and a length of 10 cm. Taking π as $\frac{22}{7}$ and the density of the material as 7·5 g/cm³, calculate the mass of the collar (i) in grams, and (ii) in kilograms.
   (b) If the dimensions given were taken at 20 °C, calculate the temperature to which the collar has to be heated in order to increase the bore diameter by 0·003 3 cm. Take the coefficient of linear expansion as 11×10^{-6} per °C.
   (c) If the specific heat capacity of the material is 400 J/kg K, calculate the amount of heat energy required for the heating in joules.

Most relevant sections: 1.3, 14.2.1, and 12.1

C.G.L.I.

5. Determine the thermal coefficient of linear expansion of the material of a rod of length 1·0 m at 20 °C which expands to 1001·28 mm at 100 °C.

Most relevant section: 14.2.1

Y.C.F.E.

6. Describe briefly one example of the way that thermal expansion of a metal may be used in engineering practice.

Most relevant section: 14.1

U.E.I.

7. If the thermal coefficient of linear expansion of copper is 0·000 017 per °C, find the change in volume of a block of copper occupying 1000 cm³, when its temperature is raised by 20 °C.

Most relevant section: 14.4.1

U.E.I.

8. Define thermal coefficient of linear expansion. In an experiment it was found that a copper rod 500 mm long increased in length by 0·68 mm when the temperature was raised by 80 °C. Find the thermal coefficient of linear expansion of copper.

Most relevant section: 14.2.1

U.E.I.

9. In an experiment to find the thermal coefficient of linear expansion, a steel bar 500 mm long was heated in a steam jacket from 15 °C to 100 °C, its extension being found to be 0·45 mm.
   (a) Calculate the thermal coefficient of linear expansion given by these results.
   (b) Make a careful sketch of a piece of apparatus suitable for conducting this experiment, showing clearly how the extension of the rod is measured accurately.
   (c) What are the likely sources of error in this experiment?

Most relevant section: 14.2.1

C.G.L.I.

10. At 15 °C a steel rod is 200 mm long and an aluminium rod is 199·95 mm long. Find the temperature at which their lengths will be the same.
    Describe briefly a device which makes use of expansion to control temperature.
    Thermal coefficient of linear expansion of steel = 11×10^{-6} per °C.
    Thermal coefficient of linear expansion of aluminium = 23×10^{-6} per °C.

Most relevant sections: 14.2.1 and 14.2.2

E.M.E.U.

11. (a) Explain briefly, with the aid of simple diagrams, the behaviour of a bimetallic strip when subjected to increasing temperature.

(b) A steel collar of mass 0·75 kg and internal diameter 49·89 mm at 20 °C is to be shrunk onto a shaft of diameter 50·00 mm at 20 °C. Calculate:

(i) the temperature to which the collar must be raised so that it may be fitted when there is a diameter clearance of 0·01 mm, and

(ii) the heat absorbed by the collar during the process. Take the specific heat capacity of steel as 0·48 kJ/kg K and the thermal coefficient of linear expansion of steel as 12×10^{-6} per °C.

Most relevant sections: 14.2.2, 14.2.1, and 12.1

U.E.I.

12. (a) Give one example of the way that thermal expansion of a metal may be

(i) an *advantage*,
(ii) a *disadvantage* in engineering practice.

(b) Explain briefly, with the aid of simple diagrams, the behaviour of a bimetallic strip when subjected to increasing temperature.

(c) The thermostat switch for an electrical immersion heater is operated by a brass rod. The rod is 300 mm long at 10 °C and the switch opens when the water temperature reaches 80 °C. Given that the thermal coefficient of linear expansion for the brass is 19×10^{-6} per °C, calculate the increase in length of the brass rod required to open the switch.

Most relevant sections: 14.1, 14.2.2, and 14.2.1

U.E.I.

13. (a) Give one example of the way that thermal expansion of a metal may be (i) an advantage, (ii) a disadvantage in engineering practice.

(b) What is meant by the 'coefficient of linear expansion' of a material?

(c) In an experiment to determine the coefficient of linear expansion of copper, a copper rod was heated in a steam jacket and the following results obtained:

original length of rod = 400·00 mm,
final length of rod   = 400·56 mm,
original temperature = 20 °C,
final temperature   = 100 °C.

Calculate the experimental value of the thermal coefficient of linear expansion of copper.

Most relevant sections: 14.1 and 14.2.1

U.E.I.

14. (a) Describe briefly, with the aid of simple diagrams, one example of how the thermal expansion of a material may be usefully employed in engineering practice.

(b) A rectangular block of steel has a mass of 2 kg and a length of 150 mm at a temperature of 20 °C. The steel has a specific heat capacity of 0·5 kJ/kg K and a thermal coefficient of linear expansion of 12×10^{-6} per °C. If the temperature of the block is raised to 270 °C find

(i) the quantity of heat energy absorbed by the block, and

(ii) the increase in its length.

Most relevant sections: 14.1, 12.1, and 14.2.1

U.E.I.

ANSWERS TO EXERCISES 14

1. 150·132 mm. **2.** 0·000 018 per °C. **3.** 500·11 mm. **4.** (a) 1650 g, 1·65 kg, (b) 70 °C, (c) 33 000 J. **5.** 0·000 016 per °C. **7.** 1·02 cm³. **8.** 0·000 017 per °C. **9.** 0·000 010 5 per °C. **10.** 35·8 °C. **11.** (i) 220·4 °C; (ii) 72 kJ. **12.** (c) 0·399 mm. **13.** (c) $17·5 \times 10^{-6}$ per °C. **14.** (b) 250 kJ, 0·45 mm.

15 Heat and gases

15.1. GASES AND VAPOURS

In the gaseous state, substances may exist either as gases or as vapours. A gas may be liquefied by means of increasing the pressure and reducing the temperature: a vapour can be condensed by increasing the pressure without lowering the temperature. At ordinary atmospheric temperatures and pressures air is an example of a gas; steam is an example of a vapour.

Gases and vapours have no definite volume at a given temperature but will always completely fill the vessel in which they are held.

15.2. PRESSURE OF A GAS

The molecules of a gas move rapidly about in all directions inside the container in which the gas is held. Constant collision between the molecules and the walls of the container produces forces distributed all over the inside of the vessel, and the amount of force per unit area is termed the pressure of the gas. Gas pressure is generally measured by means of a gauge which shows the pressure of the gas in relation to the pressure of the atmosphere. The absolute pressure of a gas is measured from perfect vacuum, so that

$$\text{absolute pressure} = \text{gauge pressure} + \text{atmospheric pressure}.$$

15.2.1. Barometric units.

Atmospheric pressure is recorded by means of a barometer in which the height of a liquid column is used as a measure of the pressure. In order that the height of the column can be kept as small as possible, the liquid employed is mercury (Hg).

Two of the pressure units at present used in barometry are, in order of preference:

(i) the millibar (mbar), equal to 100 Pa; and
(ii) the conventional millimetre of mercury (mmHg),

which may be defined as the pressure exerted by a one-millimetre column of liquid of uniform density 13 595·1 kg/m³ under standard acceleration 9·806 65 m/s² (1 mmHg = 133.322 Pa).

Although barometric height is convenient for measurement purposes, the engineer usually measures pressure in terms of force per unit area, e.g. N/m² or Pa. Let the height of the mercury column be h m, the density of mercury be 13 590 kg/m³, and the acceleration due to gravity be 9·81 m/s² then

$$\text{atmospheric pressure } p = \rho g h$$
$$= 13\,590 \times 9{\cdot}81 \times h \text{ Pa}$$
$$= 0{\cdot}1333\, h \text{ MPa}.$$

15.2.2. Pressure gauges.

Most pressure gauges indicate the difference between the atmospheric pressure and the pressure of the gas inside a container.

For *low pressures*, a U-tube gauge called a *manometer* may be used. One end of a glass tube is fitted to the vessel, the other end being open to the atmosphere (see Fig. 15.1). Some liquid, usually water or mercury, is poured into the tube, and if the gas pressure in the vessel is greater than the pressure of the atmosphere, the liquid level will be as indicated by Fig. 15.1(a). The gauge pressure will be added to the atmospheric pressure as read from a barometer.

Conversely, if the gaseous pressure inside the vessel is less than atmospheric pressure (Fig. 15.1(b)), the gauge pressure is subtracted from the barometric height.

The pressure of the atmosphere is measured as the height of a column of mercury, and since the relative density of mercury is 13·59,

then, *for a water manometer*,
  absolute pressure = (barometric height $\pm h/13{\cdot}59$) mmHg and, *for a mercury manometer*,
  absolute pressure = (barometric height $\pm h$) mmHg.

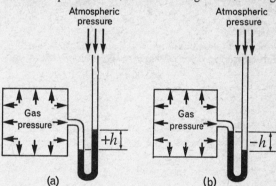

(a)
Gas pressure > Atmospheric pressure

(b)
Gas pressure < Atmospheric pressure

FIG. 15.1

Where the gaseous pressure inside a vessel is considerably greater than atmospheric pressure, the manometer is unsuitable for measuring the pressure of the gas since the tube required would be very long. Pressure gauges for indicating *high pressure*, such as in a boiler or air compressor, are generally of the *Bourdon type*, the scale of which is calibrated in units such as MPa, etc.

For extra high pressures the gauge is often graduated in 'atmospheres' where one atmosphere (atm) is approximately 0·1 MPa. Thus a pressure of 50 atm = $50 \times 0{\cdot}1$ = 5 MPa.

EXAMPLE 15.1. A water manometer connected to the bottom of a boiler chimney indicates that the pressure in the chimney is 27·2 mmHg below atmospheric pressure. If the barometer reads 760 mmHg, determine the absolute pressure inside the chimney measured in (a) mmHg and (b) kPa.

SOLUTION

(*Note:* The pressure of the flue gases is below atmospheric pressure so that the gauge pressure must be subtracted from the barometric height.)

$$\text{Absolute pressure} \atop \text{of flue gases} = \left(\text{barometric height} - \frac{h}{13\cdot59}\right)\text{mmHg}$$

$$= 760 - \frac{27\cdot2}{13\cdot59}\ \text{mmHg}$$

$$= 758\ \text{mmHg}.$$

But 1 mmHg = 133.322 Pa,

∴ absolute pressure of flue gases

$$= 758 \times 133\cdot322\ \text{Pa}$$

$$= 101\ 058\ \text{Pa}$$

$$= 101\cdot058\ \text{kPa}.$$

EXAMPLE 15.2. A Bourdon pressure gauge indicates a pressure of 400 kPa. Determine the absolute pressure, in kPa, if the barometer shows an atmospheric pressure of 980 mbar.

SOLUTION

The millibar (mbar) is equal to 100 Pa, therefore atmospheric pressure = 980 × 100 Pa

$$= 98\ 000\ \text{Pa or 98 kPa}.$$

Then

$$\text{absolute pressure} = {\text{gauge} \atop \text{pressure}} + {\text{atmospheric} \atop \text{pressure}}$$

$$= 400 + 98\ \text{kPa}$$

$$= 498\ \text{kPa}.$$

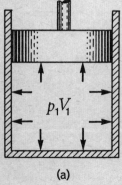 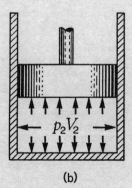

(a)   (b)

FIG. 15.2

15.3. PRESSURE AND VOLUME OF A GAS

Boyle's law. This important gas law states that *the absolute pressure p of a given mass of any gas varies inversely as the volume V, provided that the temperature remains constant.*

Consider a mass of gas under conditions p_1, V_1 (Fig. 15.2(a)), and let the gas be compressed to other conditions p_2, V_2 (Fig. 15.2(b)); then

$$\frac{p_1}{p_2} = \frac{V_2}{V_1},$$

or

$$p_1V_1 = p_2V_2.$$

Thus $pV = $ a constant.

Boyle's law is closely followed by the gases, but is not obeyed by vapours. A perfect, or ideal, gas is a theoretical gas which is imagined to obey Boyle's law.

When the conditions of a gas are changed at constant temperature, the change is said to be an 'isothermal' expansion or compression.

A graph of the change of pressure with volume has the form shown by Fig. 15.3(a). Fig. 15.3(b) shows clearly that pressure is inversely proportional to volume.

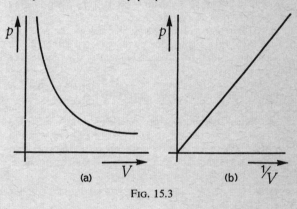

FIG. 15.3

EXAMPLE 15.3. If the gauge pressure of 2 m³ of gas is increased from 0·14 MPa to 0·175 MPa, find the new volume. The temperature remains constant and atmospheric pressure is 0·1 MPa.

Y.C.F.E.

SOLUTION

(*Note:* All pressures must be absolute.)

The temperature remains constant; applying Boyle's law:

$$p_1V_1 = p_2V_2$$

where $p_1 = 0\cdot14$ MPa (gauge) $= 0\cdot24$ MPa (abs),

$$V_1 = 2\ \text{m}^3.$$

$$p_2 = 0\cdot175\ \text{MPa (gauge)} = 0\cdot275\ \text{MPa (abs)},$$

$$V_2 = \ ?$$

Then new volume $V_2 = \dfrac{p_1 V_1}{p_2}$

$$= \frac{0.24 \times 2}{0.275} \text{ m}^3$$

$$= 1.75 \text{ m}^3.$$

15.4. VOLUME AND TEMPERATURE OF A GAS

Charles's law. This further important gas law states that *the volume V of a gas varies directly as the absolute temperature T, provided that the pressure remains constant.*

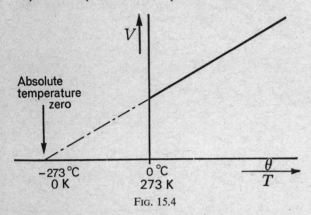

FIG. 15.4

Consider a mass of gas under conditions V_1, T_1, and let the gas be compressed or expanded to other conditions V_2, T_2; then

$$\frac{V_1}{T_1} = \frac{V_2}{T_2} = \text{a constant}.$$

A graph of the change of volume with temperature is linear as shown by Fig. 15.4, and if the line is produced backwards it always meets the temperature axis at a point known as 'absolute zero'. Absolute zero is at -273.15 °C (say -273 °C) on the temperature scale and is the temperature at which, in theory, the volume would disappear.

In dealing with gases, all temperatures must be converted to thermodynamic temperatures. Thus

$$T(\text{K}) = \theta \, (°\text{C}) + 273.$$

It will be seen that

$$0 \,°\text{C} = 273 \text{ K},$$
$$100 \,°\text{C} = 373 \text{ K}.$$

EXAMPLE 15.4. A quantity of gas has a volume of 120 cm³ at 15 °C. What will be its volume at 87 °C if the pressure is maintained constant?

U.L.C.I.

SOLUTION

(*Note:* All the temperatures must be absolute).
The pressure remains constant: applying Charles's law:

$$\frac{V_1}{T_1} = \frac{V_2}{T_2},$$

where $\quad V_1 = 120 \text{ cm}^3$,
$\qquad T_1 = 15 \,°\text{C} = 288 \text{ K}$,
$\qquad V_2 = ?$
$\qquad T_2 = 87 \,°\text{C} = 360 \text{ K}.$

Then,

$$\text{new volume } V_2 = \frac{V_1 T_2}{T_1}$$

$$= \frac{120 \times 360}{288} \text{ cm}^3$$

$$= 150 \text{ cm}^3.$$

15.5. PRESSURE, VOLUME, AND TEMPERATURE OF A GAS

In practice, whenever the conditions of a mass of gas are undergoing a change, the pressure, volume and temperature are changing simultaneously. For convenience, consider the following step-by-step process.

Fig. 15.5(a) shows a given mass of gas under conditions p_1, V_1, and T_1. Let p_1 and V_1 be changed at constant temperature T_1 until the pressure and volume become p_2 and V respectively. Then, from Boyle's law,

$$p_1 V_1 = p_2 V,$$

$$\text{so that } V = \frac{p_1 V_1}{p_2}. \tag{1}$$

The gas is now in the condition as shown by Fig. 15.5(b). Now let the temperature change from T_1 to T_2 at constant pressure p_2 so that the volume changes from V to V_2 as shown by Fig. 15.5(c). Then, from Charles's law,

$$\frac{V}{V_2} = \frac{T_1}{T_2}$$

$$\text{so that } V = \frac{V_2 T_1}{T_2}. \tag{2}$$

Equating (1) and (2),

$$\frac{p_1 V_1}{p_2} = \frac{V_2 T_1}{T_2},$$

$$\text{and } \frac{p_1 V_1}{T_1} = \frac{p_2 V_2}{T_2}.$$

Then $\qquad\qquad \dfrac{pV}{T} = \text{a constant}.$

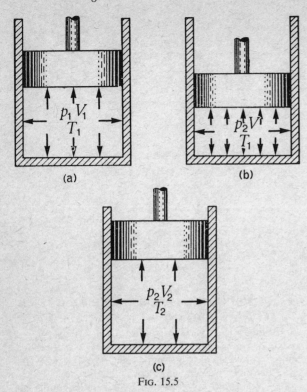

Fig. 15.5

EXAMPLE 15.5. A quantity of air occupies a volume of 4 m³ at an atmospheric pressure of 0·1 MPa absolute and a temperature of 20 °C. It is compressed to a volume of 1·5 m³ at a pressure of 0·3 MPa gauge. Calculate its temperature under the new conditions.

SOLUTION

(*Note:* All pressures and temperatures must be absolute.)

The pressure, volume, and temperature are changing simultaneously; applying the combined laws:

$$\frac{p_1 V_1}{T_1} = \frac{p_2 V_2}{T_2}$$

where $p_1 = 0\cdot1$ MPa (abs), $p_2 = 0\cdot3$ MPa (gauge)
$= 0\cdot4$ MPa (abs),

$V_1 = 4$ m³, $V_2 = 1\cdot5$ m³,

$T_1 = 20\,°C = 293$ K, $T_2 = ?$

Then new temperature $T_2 = \dfrac{p_2 V_2 T_1}{p_1 V_1}$

$$= \frac{0\cdot4 \times 1\cdot5 \times 293}{0\cdot1 \times 4}\ \text{K}$$

$$= 439\ \text{K}$$

$$= (439 - 273)\ °C$$

$$= 166\ °C.$$

SUMMARY

Absolute pressure = gauge pressure+atmospheric pressure.

Mercury barometer:

atmospheric pressure = 0·1333 h MPa, where h is the barometric height (mmHg).

Water manometer:

absolute pressure = (barometric height±h/13·59)
mmHg;

Mercury manometer:

absolute pressure = (barometric height ± h) mmHg, where h is the gauge pressure height.

Extra high pressures may be measured in atmospheres, where 1 atm = 101 325 Pa, say 0·1 MPa.

Boyle's law: The absolute pressure p of a gas varies inversely as the volume V, provided the temperature remains constant.

$$p_1 V_1 = p_2 V_2 = \text{a constant.}$$

'Isothermal' means constant temperature.

Charles's law: The volume V of a gas varies directly as the absolute temperature T, provided the pressure remains constant.

$$\frac{V_1}{T_1} = \frac{V_2}{T_2} = \text{a constant.}$$

$$T(\text{K}) = \theta\,(°\text{C}) + 273.$$

From the combined laws:

$$\frac{p_1 V_1}{T_1} = \frac{p_2 V_2}{T_2} = \text{a constant.}$$

EXERCISES 15

1. A car tyre is inflated to a gauge pressure of 180 kPa at 17 °C. Running on the tyre raises its temperature to 27 °C. Given that the atmospheric pressure is 100 kPa, what is the new gauge pressure of the tyre?

Most relevant section: 15.5

2. A sealed vessel, of volume 500 cm³, contains gas at the atmospheric pressure of 100 kPa. If the temperature of the gas is raised from 10 °C to 35 °C, without any change in volume, calculate (a) the new pressure of the gas and (b) the heat transferred to the gas. For these conditions take the density and specific heat capacity of the gas as 1·3 g/l and 700 J/kg K.

Most relevant sections: 15.5 and 12.1

3. A volume of 150 cm³ of air at atmospheric pressure (0·1 MPa abs) is pumped into a motor car tyre in which the pressure is 0·2 MPa gauge. What volume will it occupy inside the tyre assuming there has been no change of temperature?

Most relevant section: 15.3

C.G.L.I.

4. If 5 m³ of air at 15 °C enters an electric fan heater and leaves at 25 °C, find the exit volume, assuming there has been no change in pressure.

Most relevant section: 15.4

C.G.L.I.

5. A manometer connected to the inside of a tank indicates a gauge pressure of 202 mm Hg when the barometer reads 748 mm Hg.
What is the absolute pressure, in kPa, inside the tank? (1 mm Hg = 133·3 Pa).

Most relevant sections: 15.2.1 and 15.2.2

U.E.I.

6. State (a) Boyle's law, (b) Charles's law, (c) Boyle's and Charles's law combined in a single equation.
12 m³ of air at 17 °C and 0·1 MPa abs is compressed into a pressure vessel having a capacity of 1·5 m³. If the final pressure is 1·4 MPa abs find the final temperature.
If the air in the pressure vessel is now cooled to 17 °C calculate the new pressure.
Absolute zero = −273 °C (0 K).

Most relevant sections: 15.3, 15.4, and 15.5

E.M.E.U.

7. A quantity of gas occupies a volume of 5 m³ at an atmospheric pressure of 0·1 MPa absolute and a temperature of 20 °C. It is compressed to a volume of 1·8 m³ at a pressure of 0·4 MPa absolute. Calculate its temperature under the new conditions. What would be the corresponding gauge pressures?

Most relevant sections: 15.5 and 15.2

U.L.C.I.

8. What is meant by the absolute temperature of a gas? Draw graphs to show:
(a) how the volume of a gas varies with temperature when the pressure is constant, and
(b) how the volume varies with pressure when the temperature is constant.
The cylinder of an air compressor is 200 mm in diameter and 360 mm long when air is drawn in at 0·1 MPa at 15 °C.
Determine the pressure after the piston has moved 240 mm if the temperature rises to 180 °C.

Most relevant sections: 15.4 and 15.5

Y.C.F.E.

9. State Charles's law for a perfect gas.
The state of a quantity of gas enclosed in an engine cylinder at the beginning of compression is: volume 1·8 l, temperature 80 °C and pressure 0·1 MPa absolute. What will be the temperature of the gas after the volume is reduced to 0·5 l and the pressure raised to 0·75 MPa absolute?

Most relevant sections: 15.4 and 15.5

Y.C.F.E.

10. An oxygen cylinder holds 0·2 m³ of gas at a pressure of 0·5 MPa gauge and at a temperature of 10 °C.
(a) What would the pressure become if the temperature rose to 30 °C?
(b) If the temperature is maintained constant at 10 °C whilst oxygen is released slowly from the cylinder what volume measured at atmospheric pressure will be released whilst reducing the pressure in the cylinder to atmospheric?
Atmospheric pressure = 0·1 MPa.

Most relevant sections: 15.2, 15.5, and 15.3

U.E.I.

11. A test chamber of volume 1·0 m³ contains air at ambient (surrounding) conditions of 100 kPa and 17 °C. If the final condition required is 67 kPa and −17 °C calculate the volume of air at ambient (surrounding) conditions that must enter or leave the chamber.

Most relevant section: 15.5

N.C.T.E.C.

12. A pressure vessel having a volume of 4 m³ is to be charged to a pressure of 0·8 MPa. If the atmospheric pressure and temperature are 0·1 MPa and 10 °C respectively, calculate the volume of atmospheric air to be taken in during charging if the final temperature within the charged vessel is (a) 10 °C, (b) 80 °C. All pressures are absolute.

Most relevant section: 15.5

U.L.C.I.

13. State Boyle's law and sketch the graph connecting the two quantities with which this law is concerned. Air at a pressure of 0·1 MPa abs and a temperature of 17 °C is compressed to 2·4 MPa abs. Its volume is one twelfth of its original volume.

Calculate the final temperature of the air.
What would have been the final temperature and pressure if the compression had occurred in accordance with Boyle's law?

Most relevant sections: 15.3 and 15.5

U.E.I.

ANSWERS TO EXERCISES 15

1. 190 kPa. **2. (a)** 108·8 kPa, **(b)** 11·4 J. **3.** 50 cm³. **4.** 5·2 m³. **5.** 126·6 kPa. **6.** 234 °C, 0·8 MPa (abs). **7.** 149 °C, zero, 0·3 MPa (gauge). **8.** 0·473 MPa (abs). **9.** 464 °C. **10. (a)** 0·543 MPa (gauge), **(b)** 1 m³. **11.** 0·3 m³ leaves. **12. (a)** 32 m³, **(b)** 25·7 m³. **13.** 307 °C, 17 °C, 1·2 MPa (abs).

16 The electric current

16.1. NATURE OF MATTER (see also section 25.2)
All materials whether solid, liquid, or gaseous are composed of basic substances called elements.

16.1.1. Element. An element is a substance which cannot be split into different or simpler substances by any chemical means.

Typical elements are hydrogen, oxygen, iron, and copper. There are over 100 such elements and by various combinations of two or more of these every other conceivable substance is made. The substances formed by the chemical combination of two or more elements are called compounds.

16.1.2. Compound. A compound is a chemical combination of two or more elements, in fixed proportions, having properties different from those of its constituent elements.

Water, which is composed of the two elements hydrogen and oxygen, is a typical compound. The physical and chemical properties of water, a liquid at normal temperatures, are very different from those of oxygen and hydrogen which are both gases at normal temperatures.

Both compounds and elements are composed of molecules.

16.1.3. Molecule. A molecule is the smallest part of a substance which can have an independent existence.

Molecules may be further subdivided into atoms.

16.1.4. Atom. An atom is the smallest part of an element that can enter into chemical combination with an atom or atoms of other elements.

Note: Molecules of an element are composed of only one kind of atom. For example, a molecule of oxygen consists of two oxygens atoms: oxygen being an element.

Molecules of a compound are composed of two or more different kinds of atoms. Carbon dioxide, for example, is a compound whose molecules consist of one carbon atom combined with two oxygen atoms.

16.2. STRUCTURE OF THE ATOM

All atoms, with the exception of hydrogen (see Fig. 16.1), are composed of three principal types of particles, namely, electrons, protons, and neutrons. (There exist other sub-atomic particles which need not be discussed here.)

Electron. An electron is an extemely small sub-atomic particle of mass $9 \cdot 11 \times 10^{-31}$ kg and having negative electrical charge.

Proton. A proton is a very small sub-atomic particle

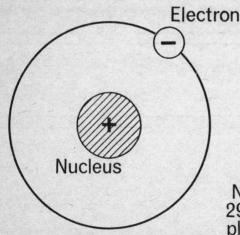

Hydrogen atom.

1 orbiting electron electrically balanced by 1 proton in the nucleus.

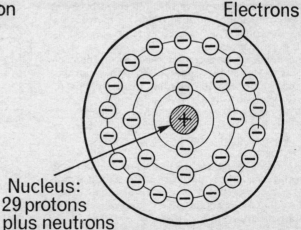

Copper atom.

The 29 negatively charged electrons electrically balance the 29 positively charged protons in the nucleus.

Fig. 16.1

of mass 1.7×10^{-27} kg (which is about 1850 times greater than that of an electron) and having positive electrical charge.

Neutron. A neutron is a very small, electrically neutral, sub-atomic particle whose mass is approximately equal to that of a proton.

The protons and neutrons form a central core or nucleus round which the electrons travel in fixed paths at very high speeds.

As the nucleus and electrons are so minute the atom consists largely of empty space.

Under normal conditions the atom is electrically neutral since the positive charge on the protons is equal and opposite to the charge on the orbiting electrons.

Atoms of two common elements hydrogen and copper, are illustrated diagrammatically in Fig. 16.1. Hydrogen is the simplest and lightest of all known elements whilst copper is one of the most important elements in electrical science.

16.3. ELECTRIC CURRENT

In certain metals, particularly in the case of copper, there is a strong tendency for the outer electron to become detached from the atom.

The loss of such an electron produces an electrically unbalanced atom having a net positive charge. This positively charged atom is called a *positive ion*.

The detached or *free* negatively charged *electrons* move at random in the spaces between the atoms frequently combining with other positive ions and then escaping again. In a quantity of material there will be at any time millions of such wandering electrons.

These wandering electrons can be influenced by some external electrical source, such as a battery, to drift in a particular direction. This directed *flow of electrons* is called an *electric current*.

The magnitude of the current depends on the number of electrons flowing and the speed at which they move.

In can be seen in Fig. 16.2 that the direction of flow of the electron current is from the negative to the positive source. This idea conflicts with a much older convention which considered an electric current to be a flow of positive charges in the reverse direction (i.e. from positive to negative).

However since most electrical phenomena can be satisfactorily explained by the latter convention it is still retained.

The directions of flow of the electron and conventional currents are illustrated in Fig. 16.3.

16.4. CONDUCTORS AND INSULATORS

In the previous section it was established that an electric current is produced by a movement of free electrons.

Materials which contain large numbers of free electrons are called *conductors*. All metals and certain non-metals, such as carbon, are good conductors of electricity. In

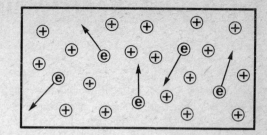

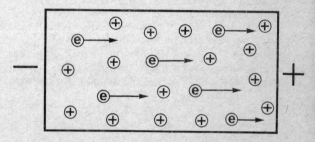

ⓔ Free electrons ⊕ Positive ions

FIG. 16.2. (a) Random movement of free electrons between positive ions. (b) Introduction of an electrical source produces electron drift in one direction

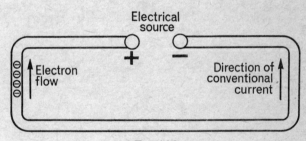

FIG. 16.3

particular, of all the metals, silver and copper show the best conducting properties.

There are, however, some materials in which all the electrons are so strongly attracted to the central nucleus that very few escape. Materials of this type in which there are few free electrons are termed non-conductors or *insulators*. Among the best insulators are rubber, bakelite and impregnated paper.

16.5. EFFECTS OF AN ELECTRIC CURRENT

The effects produced by an electric current may be grouped under three headings: *heating effects, magnetic effects, and chemical effects.*

16.5.1. Heating effects. When a current flows along a wire the wire may become hot due to some of the electrical energy being converted into heat energy.

This heating effect has wide application in both domestic and industrial fields. Familiar domestic appliances are the electric fire, electric cooker, and the electric water heater. Industrial applications include arc welding and the heating of kilns and furnaces.

16.5.2. Magnetic effects. When an electric current flows in a conductor it sets up a magnetic field round the conductor. This important discovery made by the Danish scientist *Oersted* in 1820 was the basis of a branch of electricity known as electromagnetism and made possible the development of electric motors and transformers.

16.5.3. Chemical effects. Certain fluids conduct electricity but unlike metals they undergo chemical change while the current is flowing. Fluids which conduct electricity are called electrolytes and the process of conduction through a fluid is known as electrolysis.

Important industrial applications of electrolysis include electro-plating and metal refining.

SUMMARY

An element is a substance which cannot be split up into anything simpler by a chemical change.

A compound is a substance which contains two or more elements combined in such a way that their properties are changed.

A molecule is the smallest part of an element or compound which can exist in a free state.

An atom is the smallest part of an element that can take part in a chemical change.

An electron is a minute particle of negative electricity having negligible mass. (But see section 16.2)

A proton is a very small particle of positive electricity.

A neutron is a very small electrically neutral particle.

An electric current may be defined as a flow of free electrons.

Effects produced by an electric current are heating effects, chemical effects, and magnetic effects.

EXERCISES 16

1. Atoms consist of a .... around which there are ....revolving in orbits. 'Free' electrons are........................

Most relevant sections: 16.2 and 16.3

U.E.I.

2. State which effect of an electric current is the basis for the successful operation of the following.
   (a) a filament lamp,
   (b) a primary cell,
   (c) a fuse,
   (d) an electric bell.

Most relevant section: 16.5

U.E.I.

3. State two of the effects of an electric current. Give an example of each. Sketch and describe the apparatus used to demonstrate the effects.

Most relevant section: 16.5

U.C.L.I.

4. Sketch and describe the simple structure of the atom. Make reference to the type of electric charge which exist within the atom.

Most relevant section: 16.2

U.C.L.I.

5. (a) Specify the requirements for (i) a good conductor, (ii) a good insulator.
   (b) Listed below are materials which may be used as a good conductor or good insulator. Make a list of these and state whether each material is a good conductor or good insulator.
   Copper, rubber, mica, aluminium, paper, iron, steel, steel coated aluminium, enamel, carbon, asbestos, resin, and glass.

Most relevant section: 16.4

W.J.E.C.

17 The electric circuit

17.1. QUANTITIES OF ELECTRICITY

17.1.1. The coulomb. It was established in the previous chapter that the basic unit of electrical charge is the electron. A quantity of electricity could therefore be expressed as a number of electron charges.

However, since the electron is so minute, for all practical purposes a much larger unit is used to measure quantity of electricity. This unit is called the *coulomb* and is about 6·3 trillion times the size of the charge on an electron.

1 coulomb (C) = $6·3 \times 10^{18}$ electron charges.

Note: A quantity of electricity is often referred to as 'a charge'.

17.1.2. The ampere. The magnitude of an electric current flowing in any circuit may be measured by the number of electrons passing a certain point in the circuit per second. The actual number required to give a current of 1 ampere is $6·3 \times 10^{18}$ electrons per second. See section 22.4.1. for the correct international definition of an *ampere*.

Since 1 ampere (A) = $6·3 \times 10^{18}$ electron charges per second
and 1 coulomb (C) = $6·3 \times 10^{18}$ electron charges,

1 ampere = 1 coulomb per second.

If a current of I amperes flows for t seconds and Q coulombs of electricity are transferred, then

$$I \text{ amperes} = \frac{Q}{t} \text{ coulombs per second}$$

or Q coulombs = It ampere-seconds.

Thus the coulomb may be regarded as an ampere-second.

In some engineering applications a unit much larger than the coulomb is used for quantity of electricity. This unit is termed the ampere-hour (A h).

1 ampere-hour = 3600 ampere-seconds
(1 hour = 3600 seconds),

∴ 1 ampere-hour = 3600 coulombs.

EXAMPLE 17.1. Calculate the charge in ampere-hours and in coulombs when a current of 20 A flows for 30 min.

SOLUTION
(i) Using $Q = It$,

$$Q = 20 \text{ A} \times \frac{30}{60} \text{ h},$$

∴ $Q = 10$ A h.

(ii) Again using $Q = It$,

$$Q = 20 \text{ A} \times (30 \times 60) \text{ s}$$
$$= 36\,000 \text{ C}.$$

∴ $Q = 36\,000$ C

Charge = 10 A h or 36 000 C.

17.2. RESISTANCE

All conductors, no matter how good, tend to oppose the flow of an electric current. This opposition to the flow of electrons can be attributed to the numerous collisions which occur between the free electrons and the positively charged atoms (positive ions).

A measure of this opposition to the flow of the electric current is called the *resistance* of the conductor.

The unit of resistance is the *ohm* and its symbol is Ω.

17.3. ELECTROMOTIVE FORCE AND POTENTIAL DIFFERENCE

17.3.1. Electromotive force. It has already been shown that an electric current is a flow of electrons along a conductor. However, before these electrons will flow it is necessary to introduce an electrical source, such as a battery, which will produce a difference of 'pressure' in the circuit. The electric 'force' which will produce a difference of 'pressure' is called the *electromotive force*.

Electromotive force may be defined as that electrical 'force' which is initially available for moving electrons round a circuit.

17.3.2. Potential difference. *The difference in electrical 'pressure' between any two points in a circuit is called the potential difference between the two points.* The potential difference between the terminals of a cell or battery when no current is flowing is the maximum potential difference and is equal to the electromotive force of the cell or battery.

17.3.3. Units and abbreviations. Potential difference is normally abbreviated to p.d. and has the symbol V. Electromotive force is abbreviated to e.m.f. and has the symbol E. Both e.m.f. and p.d. are measured in *volts*.

17.4. CONDUCTANCE

The reciprocal of resistance R is called conductance (symbol G). Conductance is a measure of the ease with which electricity flows. The greater the conductance of a given conductor the more easily it conducts electrical current. The unit of conductance is the siemens (S). Thus, $G = 1/R$ siemens.

It is sometimes convenient, particularly when dealing with resistors in parallel, to work in terms of conductance rather than resistance.

17.5. OHM'S LAW

George Simon Ohm, a German physicist, introduced in 1827 a law which provided a very important relationship between

potential difference (V), resistance (R), and current (I).

The potential difference (p.d.) between the ends of a conductor is directly proportional to the current flowing through it, provided that its temperature and other physical conditions do not change.

The above law may be expressed mathematically as follows:

$V \propto I$, where V is the potential difference and I the current.

An equation involving V and I may be found by the introduction of a constant k such that

$$V = kI.$$

This constant k is *defined* as the resistance R of the circuit and the equation then becomes $V = IR$.

Ohm's law may therefore be expressed by the equation

$$V = IR,$$

where *potential difference V must be expressed in volts, current I must be expressed in amperes, and resistance R must be expressed in ohms.*

Note: Multiples or submultiples of any of the above units must be expressed as a decimal of these units before substitution in the equation $V = IR$.

EXAMPLE 17.2. A car headlamp bulb takes a current of 3 A from a 12-V battery. Calculate the resistance of the bulb filament.

SOLUTION

Using $V = IR$ and transposing for R,
$$R = \frac{V}{I}.$$
$V = 12$ volts and $I = 3$ amperes,
$$\therefore R = \frac{12}{3} = 4 \text{ ohms}.$$

Resistance of lamp filament $= 4\,\Omega$.

EXAMPLE 17.3. A current of 200 mA flows through a resistance of 400 Ω. What is the p.d. across this resistance?

SOLUTION

Before using the formula $V = IR$ the 200 mA current must be expressed as a decimal of 1 A.

$$200 \text{ mA} = \frac{200}{1000} \text{ A} = 0.2 \text{ A}.$$
Using $V = IR$,
$$V = 0.2 \times 400$$
$$= 80 \text{ volts}.$$

Potential difference across resistance equals 80 V.

17.6. MEASUREMENT OF RESISTANCE BY THE AMMETER/VOLTMETER METHOD

The ammeter. An instrument for measuring the electric current flowing in a circuit is called an ammeter. The ammeter is connected *into* the circuit so as to measure the current flowing *through* it.

The voltmeter. An instrument for measuring the potential difference between any two points in a circuit is called a voltmeter. The voltmeter is connected *across* two points in the main circuit so as to measure the p.d. *between* the points.

(*Note:* A description of ammeters and voltmeters is given in Chapter 24.)

17.6.1. Ammeter/voltmeter method. The Ohm's law equation $V = IR$ may be transposed to give

$$R = \frac{V}{I}.$$

Hence the value of an unknown resistance may be found by measuring

(1) the current I passing through it,
(2) the potential difference V across it,

and then substituting these values in the above equation.

The connections for the test are shown in Fig. 17.1 in which

R is the unknown resistance,
E is the electrical supply,
S is a switch,
A is an ammeter measuring the current through R,
V is a voltmeter measuring the p.d. across R and

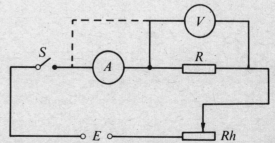

FIG. 17.1. Measurement of resistance by ammeter/voltmeter method

Rh is a variable resistance known as a rheostat. This is used for adjusting the current in the circuit to a suitable value.

The value of the unknown resistance R is then obtained by using the equation $R = \dfrac{V}{I}$.

Resistance R ohms $= \dfrac{\textbf{voltmeter reading (volts)}}{\textbf{ammeter reading (amperes)}}.$

17.7. RESISTANCES IN SERIES

The various components forming a circuit may be connected together by two fundamental methods of connection, namely, in series or in parallel.

Resistances are in *series* when they are connected end to end in such a way that they form *one path only* for the current.

Fig. 17.2 shows part of a circuit in which three resistances R_1, R_2, and R_3 are connected in series.

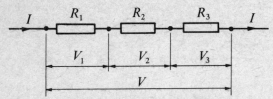

FIG. 17.2. Resistances in series

The current flowing through each resistance equals I since there is no alternative route for it to travel.

The voltage drops across each of the resistances R_1, R_2, and R_3 are V_1, V_2, and V_3 respectively.

Now the total voltage drop V across all three resistances must equal the sum of the individual voltage drops,

$$\therefore V = V_1 + V_2 + V_3.$$

Let R be a resistance equivalent to the three resistances in series. Then the current passing through R would be I and the p.d. across it would be V.

Applying Ohm's law to this equivalent circuit gives

$$V = IR. \qquad (1)$$

Now applying Ohm's law to each individual resistance gives

$$V_1 = IR_1,$$
$$V_2 = IR_2,$$
$$V_3 = IR_3.$$
$$\therefore V_1 + V_2 + V_3 = IR_1 + IR_2 + IR_3 \qquad (2)$$
$$\text{or } V_1 + V_2 + V_3 = I(R_1 + R_2 + R_3).$$
$$\text{But } V_1 + V_2 + V_3 = V = IR.$$

Replacing $V_1 + V_2 + V_3$ by IR, eqn (2) becomes

$$IR = I(R_1 + R_2 + R_3).$$
$$\therefore R = R_1 + R_2 + R_3.$$

In words, for resistances in series, *the total resistance equals the sum of the individual resistances.*

EXAMPLE 17.4. A current of 1·4 A flows through two resistances of 7 Ω and 3 Ω connected in series. Find the p.d. across each resistance and the supply voltage.

SOLUTION

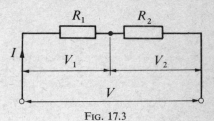

FIG. 17.3

Fig. 17.3 shows the circuit in which $I = 1\cdot4$ A, $R_1 = 3\,\Omega$, $R_2 = 7\,\Omega$, p.d. across $R_1 = V_1$, p.d. across $R_2 = V_2$ and voltage supply $= V$.

For resistances in series $R = R_1 + R_2$,

$$\therefore \text{ equivalent resistance} = 7 + 3\,\Omega = 10\,\Omega.$$

Applying Ohm's law, $V = IR$,

$$V = 1\cdot4 \times 10 = 14 \text{ volts.}$$

Voltage supply $= 14$ V.

To find p.d. across V_1,

using $V_1 = IR_1$,
$$V_1 = 1\cdot4 \times 3 = 4\cdot2 \text{ volts.}$$

P.D. across 3 Ω equals 4·2 V.
To find p.d. across V_2,

using $V_2 = IR_2$,
$$V_2 = 1\cdot4 \times 7 = 9\cdot8 \text{ volts.}$$

P.D. across 7 Ω equals 9·8 V.
Note: $V = V_1 + V_2 = 4\cdot2 + 9\cdot8 = 14$ V.

17.8. RESISTANCES IN PARALLEL

Resistances are said to be connected in *parallel* when they form *separate branches* of a circuit, the total current being divided between the individual resistances.

Fig. 17.4 shows part of a circuit in which three resistances R_1, R_2, and R_3 are connected in parallel.

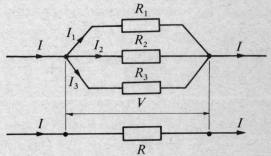

FIG. 17.4. Resistances in parallel

Let the total current be I and the currents passing through the resistances R_1, R_2, and R_3 be I_1, I_2, and I_3 respectively. Now $I = I_1+I_2+I_3$, since the current is not stored up or destroyed in passing through the resistances.

Also the p.d. V across the resistances is the same.

Applying Ohm's law to the individual branch resistances,

$$I_1 = \frac{V}{R_1},$$

$$I_2 = \frac{V}{R_2},$$

$$I_3 = \frac{V}{R_3}.$$

Adding the equations, $I_1+I_2+I_3 = \dfrac{V}{R_1}+\dfrac{V}{R_2}+\dfrac{V}{R_3}$ (1)

Now consider a single resistance R which is the exact equivalent of the three in parallel. Then the p.d. across this resistance R would be V when it was carrying a total current I.

Applying Ohm's law to this equivalent resistance R,

$$I = \frac{V}{R}.$$ (2)

But $I = I_1+I_2+I_3 = \dfrac{V}{R}.$

Replacing $I_1+I_2+I_3$ by $\dfrac{V}{R}$ in eqn (1)

$$\frac{V}{R} = \frac{V}{R_1}+\frac{V}{R_2}+\frac{V}{R_3}.$$

Dividing through by V which is common, the equation becomes

$$\frac{1}{R} = \frac{1}{R_1} + \frac{1}{R_2}\quad\frac{1}{R_3}.$$

In words, for resistances in parallel *the reciprocal of the equivalent resistance equals the sum of the reciprocals of the individual resistances.*

In section 17.4 it was stated that the reciprocal of resistance R is called conductance G. Thus for conductances G_1, G_2, G_3, etc., in parallel, the total conductance G is given by:

$$G = G_1+G_2+G_3 \text{ (siemens).}$$

EXAMPLE 17.5. Calculate the total resistance when the following resistances are connected in parallel:

(a) $2\,\Omega$, $4\,\Omega$
(b) $9\,\Omega$, $18\,\Omega$, $36\,\Omega$.

SOLUTION
(a)
Let R be the equivalent resistance of the $2\,\Omega$ and $4\,\Omega$ resistances in parallel, then

$$\frac{1}{R} = \frac{1}{2}+\frac{1}{4},$$

$$\frac{1}{R} = \frac{2+1}{4},$$

$$\frac{1}{R} = \frac{3}{4}\left(G = \frac{3}{4}\text{ siemens}\right).$$

$$\therefore R = \frac{4}{3}\,\Omega$$

$$= 1\cdot33\,\Omega.$$

The equivalent resistance $= 1\cdot33\,\Omega$.

(b)
Let R be the equivalent resistance of the $9\,\Omega$, $18\,\Omega$, and $36\,\Omega$ resistances in parallel, then

$$\frac{1}{R} = \frac{1}{9}+\frac{1}{18}+\frac{1}{36},$$

$$\frac{1}{R} = \frac{4+2+1}{36} = \frac{7}{36}\left(G = \frac{7}{36}\text{ siemens}\right).$$

$$\therefore R = \frac{36}{7} = 5\cdot14\,\Omega.$$

The equivalent resistance is $5.14\,\Omega$.

(It will be seen that the equivalent resistance in each case is less than the smallest individual resistance in the respective group. This fact should be remembered.)

EXAMPLE 17.6. Two resistances of $100\,\Omega$ and $40\,\Omega$ are connected in parallel across a 240-V supply.

Calculate (1) the combined resistance,
(2) the total circuit current, and
(3) the current through each resistance.

SOLUTION

Fig. 17.5 shows the circuit in which $R_1 = 40\,\Omega$, $R_2 = 100\,\Omega$, $I =$ main circuit current, $I_1 =$ current through R_1, $I_2 =$ current through R_2, and $V = $ 240-V supply.

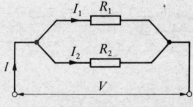

FIG. 17.5

(1) Let the combined resistance of R_1 and $R_2 = R$,

then $\frac{1}{R} = \frac{1}{40} + \frac{1}{100}$

$\frac{1}{R} = \frac{10+4}{400} = \frac{14}{400}$,

$\frac{1}{R} = \frac{1}{28 \cdot 6}$.

$\therefore R = 28 \cdot 6\,\Omega.$

The combined resistance $= 28 \cdot 6\,\Omega.$

(2) To find circuit current,

transpose $V = IR$,

$\therefore I = \frac{V}{R}$,

$I = \frac{240}{28 \cdot 6} = 8 \cdot 4\,\text{A}.$

The current $= 8 \cdot 4\,\text{A}.$

(3) To find current through R_1,

$I_1 = \frac{V}{R_1} = \frac{240}{40} = 6\,\text{A}.$

The current $= 6\,\text{A}.$
To find current through R_2,

$I_2 = \frac{V}{R_2} = \frac{240}{100} = 2 \cdot 4\,\text{A}.$

The current $= 2 \cdot 4\,\text{A}.$
Note: $I = I_1 + I_2 = 2 \cdot 4 + 6 = 8 \cdot 4\,\text{A}.$

EXAMPLE 17.7. Three resistors of $10\,\Omega$, $20\,\Omega$, and $20\,\Omega$ are connected in parallel. Two other resistors of $15\,\Omega$ and $30\,\Omega$ are also connected in parallel. The two groups of resistors are then connected in series with a 240-V electric supply.

Make a diagram of the circuit and determine the total current taken from the supply.

U.L.C.I.

SOLUTION

See Fig. 17.6.

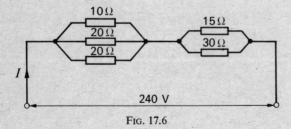

FIG. 17.6

Let $R_1\,\Omega$ be the combined resistance of the $10\,\Omega$, $20\,\Omega$, and $20\,\Omega$ resistances in parallel,

then $\frac{1}{R_1} = \frac{1}{10} + \frac{1}{20} + \frac{1}{20}$,

$\frac{1}{R_1} = \frac{2+1+1}{20} = \frac{4}{20}$,

$\therefore R_1 = 5\,\Omega.$

The combined resistance in parallel is $5\,\Omega.$
Let R_2 be the combined resistance of the $15\,\Omega$ and $30\,\Omega$ resistances,

then $\frac{1}{R_2} = \frac{1}{15} + \frac{1}{30}$,

$\frac{1}{R_2} = \frac{2+1}{30} = \frac{3}{30}$,

$\therefore R_2 = \frac{30}{3} = 10\,\Omega.$

The combined resistance is $10\,\Omega.$
If R is the total resistance of the circuit, then

$R = R_1 + R_2$ since R_1 and R_2 are in series,

$\therefore R = 5 + 10 = 15.$

Total circuit resistance $= 15\,\Omega.$

To find total current,

$I = \frac{V}{R}$

$= \frac{240}{15} = 16\,\text{A},$

$\therefore$ total current $= 16\,\text{A}.$

17.9. RESISTIVITY

Three of the factors affecting the resistance of a conductor are:

(1) its *length l*;
(2) its *area* of cross-section A;
(3) a property of the material of the conductor known as its *resistivity*.

17.9.1. Length *l*. Consider two pieces of the same conductor, one piece being twice the length of the other. Then the longer conductor may be considered as made up of two of the shorter conductors joined end to end. This is equivalent to connecting them in series and from the result obtained earlier in the chapter for series connections, the total resistance of the longer conductor is equal to twice the resistance of the shorter conductor.

It may be assumed therefore that the resistance R of a conductor is directly proportional to its length *l*, or $R \propto l$.

17.9.2. Area of cross-section A. The effect of the cross-sectional area of a conductor on resistance may be seen by considering two exactly similar conductors connected in parallel each having resistance $R\,\Omega$, then

$$\frac{1}{\text{combined resistance}} = \frac{1}{R} + \frac{1}{R} = \frac{2}{R},$$

$\therefore$ combined resistance $= \dfrac{R}{2}.$

This calculation shows that the resistance is halved by placing the two conductors in parallel.

However, placing the two conductors in parallel may be considered as equivalent to doubling the cross-sectional area of one of them. Hence by doubling the area of cross-section of a conductor its resistance is halved.

In general the larger the area of cross-section the smaller will be the resistance offered.

Resistance is inversely proportional to the cross-sectional area,

$$\text{or } R \propto \frac{1}{A}.$$

17.9.3. Resistivity.
Combining $R \propto l$ and $R \propto \dfrac{1}{A}$ gives

$$R \propto \frac{l}{A}.$$

The introduction of a constant to this statement provides an equation of the form

$$R = \frac{l}{A} \times \text{constant}.$$

This constant is called the 'resistivity' of the conductor and its value depends upon the material of the conductor.

Resistivity is denoted by the symbol ρ.

$$\therefore R = \frac{\rho l}{A}.$$

The resistivity of a material may be defined as the resistance offered by a unit cube of that material.

17.9.4. Units of resistivity.
Transposing the equation,

$$R = \frac{\rho l}{A} \text{ for } \rho,$$

$$\rho = \frac{RA}{l}.$$

If A is expressed in square metres (m^2), R in ohms (Ω) and l in metres (m), then

$$\rho = \frac{R\,(\Omega) \times A\,(m^2)}{l\,(m)}$$

and ρ is expressed in ohm-metres (Ω m).

Note: In practice it is found more convenient to give the resistivity in micro-ohm-millimetres ($\mu\Omega$ mm). For example, the resistivity of nickel is $0.000\,000\,136\,\Omega$ m which is better written as $136\,\mu\Omega$ mm.

The following table gives the resistivity at $20\,°C$, of a number of materials.

TABLE 17.1

Material	Resistivity ($\mu\Omega$ mm)
Aluminium, hard-drawn	28.0
Carbon, graphitic	46 000
Copper, standard annealed	17.2
Iron, wrought (electric)	107
Manganin (Cu, Mn, Ni)	480
Nichrome (Ni, Cr)	1090
Nickel	136
Silver, annealed	15.8
Silver, German (Cu, Ni, Zn)	344

Note: The value of resistivity varies with alterations in temperature.

From the table it can be seen that the best conductor is silver. However, its application in electrical engineering is very limited due to its high cost.

Copper is the next best conductor of electricity and since there are large quantities of it available it is widely used in the electrical industry. Some of its uses include cores of electrical wires and cables, windings on electric motors and generators and connecting conductors (busbars) on switchboards.

Aluminium, although a poorer conductor than copper is much lighter in weight. It is, therefore, used extensively on overhead transmission lines. Aluminium is also widely used for busbar systems on switchboards.

EXAMPLE 17.8. Calculate the resistance of 20 m of copper wire having a cross-sectional area of $0.05\,mm^2$. Take the resistivity of copper as $17\,\mu\Omega$ mm.

SOLUTION

Area of cross-section A of wire $= 0.05\,mm^2$, length of wire, $l = 20\,m \times 1000 = 20\,000$ mm.

$$\rho = 17\,\mu\Omega\text{ mm} = \frac{17}{10^6}\,\Omega\text{ mm}$$

$$\left(1\mu\Omega\text{ mm} = \frac{1\,\Omega\text{ mm}}{10^6}\right).$$

Using $R = \rho \dfrac{l}{A}$,

then $R = \dfrac{17 \times 20\,000}{10^6 \times 0.05},$

$\therefore R = \dfrac{17 \times 0.02}{0.05},$

$R = 6.8.$

Resistance of 20 m of the copper wire equals $6.8\,\Omega$.

EXAMPLE 17.9. The element of an immersion heater is to be made from nichrome wire of diameter 0·25 mm. Calculate the length of wire required in order that its resistance is 90 Ω.

Take the resistivity of nichrome as 1000 $\mu\Omega$ mm.

SOLUTION

Using $R = \rho \dfrac{l}{A}$ and transposing for l,

$$l = \frac{RA}{\rho}, \text{ where}$$

$$R = 90,$$

$$A = \frac{\pi(0·25)^2}{4}, \text{ (area of circular cross-}$$

$$\text{section} = \frac{\pi d^2}{4} \text{ and take } \pi = 22/7).$$

$$\rho = 1000 \ (\mu\Omega \text{ mm}) = \frac{1000}{10^6} (\Omega \text{ mm})$$

$$\left(1 \ \mu\Omega = \frac{1}{10^6} \Omega\right),$$

$$l = \frac{90 \times \pi \ (0·25)^2}{\dfrac{1000}{10^6} \times 4},$$

$$\therefore l = \frac{90 \times 22 \times 0·0625 \times 10^6}{7 \times 1000 \times 4}$$

$$l = 4420 \text{ approximately.}$$

Length required = 4420 mm or 4·42 m.

EXAMPLE 17.10. A copper conductor 550 m long is to carry a current of 400 A the voltage drop being 10 V. Calculate the diameter of the conductor.

At the working temperature, resistivity of copper = 17 $\mu\Omega$ mm. Take π as 22/7.

SOLUTION

Using Ohm's law to find the resistance,

$$R = \frac{V}{I},$$

$$\therefore R = \frac{10}{400} = \frac{1}{40}.$$

The resistance $= \dfrac{1}{40} \Omega$.

To find the diameter of the conductor,

$$A = \rho \frac{l}{R},$$

$$\rho = 17 \ (\mu\Omega \text{ mm}) = \frac{17}{10^6} (\Omega \text{ mm}),$$

$$l = 550 \ (\text{m}) = 550 \times 10^3 \ (\text{mm}),$$

$$R = \frac{1}{40},$$

$$\therefore A = \frac{17}{10^6} \times \frac{550 \times 10^3}{\dfrac{1}{40}},$$

$$A = \frac{17 \times 55 \times 4}{10} = 17 \times 22.$$

But $A = \pi \dfrac{D^2}{4}$,

$$\therefore D^2 = \frac{4 \times 17 \times 22}{22/7},$$

$$D^2 = 476,$$

$$\therefore D = 21·82.$$

Diameter required = 21·82 mm.

SUMMARY

Quantity of electricity is measured in coulombs.
1 coulomb (C) = $6·3 \times 10^{18}$ electrons approximately.
A coulomb may be regarded as an ampere-second.
1 ampere-hour (A h) = 3600 ampere-seconds,
∴ 1 ampere-hour = 3600 coulombs.

The unit of current (I) is the ampere (A).

1 ampere = 1 coulomb per second.
Q coulomb = It ampere-seconds.

The unit of resistance is the ohm: symbol Ω (omega).

Electromotive force (e.m.f.) may be defined as that electrical force which is initially available for moving electrons round a circuit.

Potential difference (p.d.) is the difference in electrical pressure between two points in a circuit.

Electromotive force and potential difference are both measured in volts (V).

Ohm's law: The p.d. between the ends of a conductor is directly proportional to the current flowing through it, provided its temperature remains constant.

Ohm's law equation,

$$V \text{ (volts)} = I \text{ (amperes)} \times R \text{ (ohms)}.$$

For resistances in series,

$$R = R_1 + R_2 + R_3 + \text{etc.}$$

The equivalent resistance R of a circuit is equal to the sum of the individual resistances, R_1, R_2, R_3.

For resistances in parallel,

$$\frac{1}{R} = \frac{1}{R_1} + \frac{1}{R_2} + \frac{1}{R_3} + \text{etc.}$$

The reciprocal of the equivalent resistance equals the sum of the reciprocals of the individual resistances.

Note: The reciprocal of R is $\dfrac{1}{R}$ i.e. the conductance G.

Conductance is measured in siemens (S).

Resistivity ρ is defined as the resistance offered by a unit cube of material.

The resistance, resistivity, cross-sectional area, and length of a conductor are connected by the equation

$$R = \rho\frac{l}{A}.$$

The preferred unit of resistivity is the micro-ohm-millimetre ($\mu\Omega$ mm).

EXERCISES 17

1. If 4 A h of electricity flow in an electrical circuit in 10 min, what current would be indicated on an ammeter connected in the circuit?

Most relevant section: 17.1.1

U.E.I.

2. Distinguish between the terms electromotive force and potential difference.

Most relevant section: 17.3

U.L.C.I.

3. A potential difference of 12 V is applied to a resistor of 3 Ω.

(a) The current flowing is...............................
(b) The quantity of electricity passing through the resistor in 2 min is..............................

Most relevant sections: 17.5 and 17.1.1

U.E.I.

4. A coil consists of 1000 turns of aluminium wire of cross-sectional area 0·7 mm². The mean length of each turn is 100 mm. The resistivity of aluminium is 28 $\mu\Omega$ mm. Calculate:

(a) the resistance of the coil;
(b) the potential difference across the coil if a current of 5 A is passed through the coil.

Most relevant sections: 17.9.3 and 17.5

U.L.C.I.

5. Three coils P, Q, and R are connected to a 40-V supply as shown in Fig. 17.7. The resistances of coils P and Q are 12 Ω and 15 Ω respectively, and the current taken from the supply is 4 A.

(a) Draw the circuit diagram and, in addition, show how an ammeter and a voltmeter should be connected to determine the resistance of coil R.

(b) Calculate the resistance of coil R and the p.d. across it.

Most relevant sections: 17.7, 17.8, 17.5, and 17.6.1

U.E.I.

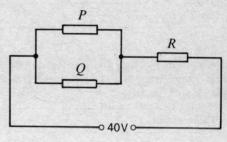

FIG. 17.7

6. If the resistance of a 10 mm cube of copper is 1·72 $\mu\Omega$, calculate

(a) the resistance of a 5 mm square section block, 15 mm long,
(b) the resistance of a 20 mm diameter rod, 3 m long.

Most relevant section: 17.9.3

U.E.I.

7. Three resistors are in series across a supply of 235 V and a 5 A current is flowing. If two of the resistors are 10 Ω and 25 Ω,

(a) what is the resistance of the third resistor, and
(b) what voltage drop would occur across the 25 Ω resistor?
(c) If the 10 Ω and 25 Ω resistors were connected in parallel, what fourth resistor in series with the third would be required to maintain the same total resistance?

Most relevant sections: 17.5, 17.7, and 17.8

U.E.I.

8. An ammeter of resistance 20 Ω is connected in series with a coil whose resistance is to be measured. When a current of 3 A passes through the coil, the potential difference across both coil and ammeter is 240 V.

What is the true resistance of the resistor? What is the percentage error in calculating the resistance directly from the meter readings?

Most relevant sections: 17.5 and 17.7

U.L.C.I.

9. What is meant by the equivalent resistance of a circuit? Find the equivalent resistance of three resistors of value $20\,\Omega$, $60\,\Omega$, and $30\,\Omega$ when connected in parallel.

Most relevant section: 17.8

Y.C.F.E.

10. A piece of wire 5 m long and $1 \cdot 0\,\text{mm}^2$ cross-sectional area has a resistance of $0 \cdot 14\,\Omega$. What is its resistivity? State the units clearly.

Most relevant section: 17.9.3

C.G.L.I.

11. Calculate the resistance of 1 km of wire having a cross-sectional area of $20\,\text{mm}^2$ and a resistivity of $28\,\mu\Omega$ mm.

Most relevant section: 17.9.3

N.C.T.E.C.

12. A circuit supplied at 150 V has two groups of parallel resistors in series. The first group contains three resistors of $2\,\Omega$, $4\,\Omega$, and $12\,\Omega$ in parallel and the second group contains two resistors of $2\,\Omega$ and $6\,\Omega$ in parallel. Determine (a) the total current flowing and (b) the current through the $4\,\Omega$ resistor.

Most relevant sections: 17.7, 17.8, and 17.5

Y.C.F.E.

13. (a) State how the resistance of electrical conductors of the same material varies:

(i) with length,
(ii) with cross-sectional area.
(b) What length of manganin wire, having a uniform cross-sectional area of $2 \cdot 5\,\text{mm}^2$, will be required to provide a resistance of $2 \cdot 5\,\Omega$?
The resistivity of manganin may be taken as $480\,\mu\Omega$ mm.

Most relevant section: 17.9.3

U.E.I.

14. Fig. 17.8 shows resistors *A*, *B*, and *C* in parallel with each other, and then connected in series with another resistor *D* across a supply of 24 V. The resistors have the following values: $A\ 8\,\Omega$, $B\ 12\,\Omega$, $C\ 16\,\Omega$, and $D\ 4\,\Omega$. Find:
(a) the total current taken from the supply;
(b) the potential difference across *D*;
(c) the current in each of the resistors *A*, *B*, and *C*.
Show how to check results (c) against result (a).

Most relevant sections: 17.8, 17.7, and 17.5

U.E.I.

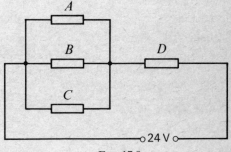

FIG. 17.8

15. A circuit consisting of two resistors R_1 and R_2 joined in parallel, is connected in series to a third resistor R_3 across a 24-V d.c. supply. The resistors have the following values: $R_1 = 2\,\Omega$, $R_2 = 6\,\Omega$, and $R_3 = 4 \cdot 5\,\Omega$.

Sketch the circuit diagram and calculate the current flowing through each resistor and the p.d. across resistor R_3.

Most relevant sections: 17.5, 17.7, and 17.8

U.E.I.

16. A resistor is made from 4 m of nichrome wire having a cross-sectional area of $0 \cdot 1\,\text{mm}^2$. Take the resistivity of nichrome as $1000\,\mu\Omega$ mm and calculate the resistance of the resistor.

A second resistor is made from 4 m of nichrome wire having a cross-sectional area of $0 \cdot 2\,\text{mm}^2$ and the two resistors are then connected in parallel to a 100-V d.c. supply. Determine the combined resistance of the resistors and the current taken from the supply.

Most relevant sections: 17.8 and 17.9

U.E.I.

17. (a) Name, in each case, one material whose electrical resistance:

(i) increases with a rise in temperature,
(ii) decreases with a rise in temperature, and
(iii) remains practically constant with a rise in temperature.

(b) A resistor, of resistance $80\,\Omega$, is made from 4 m of nichrome wire having a cross-sectional area of $0\cdot05\ mm^2$. Calculate the resistivity of the nichrome.

Most relevant sections: 17.9 and 18.1

U.E.I.

18. Three equal resistors, when connected in parallel across a 120-V d.c. supply, draw a total current of 30 A. Calculate the value of each resistor.
Draw a circuit diagram to show a fourth resistor arranged so that the total current will be reduced to 20 A. Calculate the value of this resistor.

Most relevant sections: 17.5, 17.7, and 17.8

U.E.I.

19. Three resistors are connected, as shown in Fig. 17.9, to a 24-V d.c. supply.
The resistance of resistor A is $6\,\Omega$.
The resistance of resistor B is $4\,\Omega$.

The resistance of resistor C can be varied from zero to $8\,\Omega$.
Calculate:
(a) the maximum supply current,
(b) the minimum supply current, and
(c) the maximum quantity of electricity taken from the supply in 30 s.

Most relevant sections: 17.1, 17.5, 17.7, and 17.8

U.E.I.

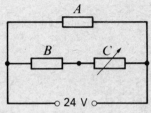

FIG. 17.9

ANSWERS TO EXERCISES 17

1. 24 A. **3.** (a) 4 A, (b) 480 C. **4.** (a) $4\,\Omega$, (b) 20 V. **5.** $3\cdot33\,\Omega$, $13\cdot33$ V. **6.** (a) $10\cdot32\ \mu\Omega$, (b) $164\ \mu\Omega$. **7.** (a) $12\,\Omega$, (b) 125 V, (c) $27\cdot86\,\Omega$. **8.** $60\,\Omega$, $33\cdot3$ per cent. **9.** $10\,\Omega$. **10.** $28\ \mu\Omega$ mm. **11.** $1\cdot4\,\Omega$. **12.** (a) $55\cdot5$ A, (b) $16\cdot65$ A. **13.** 13 m. **14.** (a) $3\cdot12$ A, (b) $12\cdot48$ V, (c) $1\cdot44$ A, $0\cdot96$ A, $0\cdot72$ A. **15.** 3 A, 1 A, 4 A, 18 V. **16.** $40\,\Omega$, $13\cdot33\,\Omega$, $7\cdot5$ A. **17.** $1000\ \mu\Omega$ mm. **18.** $12\,\Omega$, $2\,\Omega$. **19.** (a) 10 A, (b) 6 A, (c) 300 C.

18 Resistance and temperature

18.1. TEMPERATURE CHANGE AND RESISTANCE

The resistance of most materials is affected to some extent by variations of temperature and the nature of the effect is dependent upon the composition, physical condition, etc., of the material concerned.

All pure metals, such as aluminium, copper, and nickel, and certain alloys, such as brass and bronze, show a marked increase in resistance as the temperature is increased. Special alloys, known as 'resistance alloys', e.g. manganin (Cu, Mn, Ni), exhibit very little or no increase, while with materials, such as paper, cotton, and rubber, the resistance decreases quite rapidly as the temperature is increased. The resistance of carbon and electrolytes also falls with a rise in temperature.

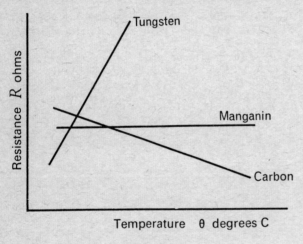

FIG. 18.1

In general, an increase in the temperature of a conductor results in an increase in its resistance (and resistivity): the resistance of an insulator diminishes as the temperature rises. This has an evident bearing on the allowable working temperature ranges of electrical machinery and plant since any change of resistance produces a variation in the current which flows.

With most materials and provided the range of temperatures is reasonably small, the resistance changes fairly regularly, so that a graph of resistance against temperature is approximately a straight line. The graphs shown in Fig. 18.1 illustrate the dissimilar ways in which tungsten, manganin, and carbon behave with changes in temperature.

18.2. TEMPERATURE COEFFICIENT OF RESISTANCE

If the resistance of a piece of material is measured at various temperatures, a graph of resistance against temperature may be drawn and extended backwards to meet the resistance axis at zero temperature as shown in Fig. 18.2. It will be seen that for an increase in temperature of 1 °C the resistance changes by a small amount r. The ratio r/R_0 is known as the 'temperature coefficient of resistance' (symbol α) and may be defined as follows:

The temperature coefficient of resistance of a material is the change in resistance of 1 ohm at 0 °C per degree temperature rise.

The unit of temperature coefficient of resistance is ohm per ohm per degree Celsius, more simply, per °C (or / °C).

Let R_0 = resistance at 0 °C,

r = increase in resistance due to temperature rise of 1 °C.

Then R_0+r = resistance at 1 °C,

and $R_0+r\theta$ = resistance at θ °C.

But $\alpha_0 = r/R_0$ so that $r = R_0\,\alpha_0$.

$$\therefore\ R_\theta = R_0+R_0\,\alpha_0\,\theta,$$
$$R_\theta = R_0(1+\alpha_0\,\theta). \tag{1}$$

In practice it is generally not possible to measure the resistance at 0 °C, and an expression which does not contain R_0 is required.

Let R_0 = resistance at 0 °C,

R_1 = initial resistance at some temperature θ_1 °C,

R_2 = final resistance at some other temperature θ_2 °C.

Then from eqn (1):

$$R_1 = R_0(1+\alpha_0\,\theta_1) \tag{2}$$
$$\text{and } R_2 = R_0(1+\alpha_0\,\theta_2). \tag{3}$$

Dividing (2) by (3):

$$\frac{R_1}{R_2} = \frac{1+\alpha_0\,\theta_1}{1+\alpha_0\,\theta_2}. \tag{4}$$

The value of α is negative for materials which show a decrease in resistance as the temperature is increased.

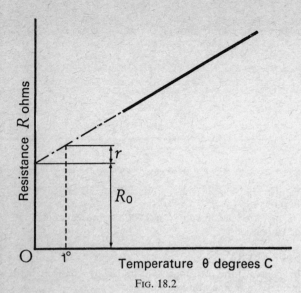

FIG. 18.2

The resistance at $0 \,°C = 0.252 \; \Omega$.

At $25 \,°C$, $R_\theta = R_0(1 + \alpha_0 \, \theta)$

$\qquad = 0.252 \, (1 + 0.0043 \times 25)$

$\qquad = 0.252 \times 1.108$

$\qquad = 0.279$.

The resistance at $\theta \,°C = 0.279 \; \Omega$.

Then, from $I = \dfrac{V}{R}$,

$\quad V = IR$

$\qquad = 30 \times 0.279$

$\qquad = 8.37$ volts.

Voltage drop $= 8.37$ V.

EXAMPLE 18.2. A relay coil takes a current of 0.5 A when at a temperature of $20 \,°C$ and connected to a 240-V supply. If the minimum operating current of the relay is 0.4 A, calculate the temperature above which the relay will fail to operate. The temperature coefficient of resistance of the coil material is $0.004/ \,°C$ from and at $0 \,°C$.

SOLUTION

When current $I = 0.5$ A:

$\quad$ resistance of coil $= \dfrac{240}{0.5} = 480 \; \Omega$.

When current $I = 0.4$ A:

$\quad$ resistance of coil $= \dfrac{240}{0.4} = 600 \; \Omega$.

$\qquad$ From $\dfrac{R_1}{R_2} = \dfrac{1 + \alpha_0 \, \theta_1}{1 + \alpha_0 \, \theta_2}$,

$\qquad \dfrac{480}{600} = \dfrac{1 + 0.004 \times 20}{1 + 0.004 \, \theta_2}$,

$0.8 \, (1 + 0.004 \, \theta_2) = 1.08$,

$\qquad 0.004 \, \theta_2 = \dfrac{1.08}{0.8} - 1$

$\qquad\qquad = 0.35$,

$\qquad$ and $\theta_2 = 87.5 \,°C$.

The relay will not operate above a temperature of $87.5 \,°C$.

18.3. REFERENCE TEMPERATURE

The value of the temperature coefficient of resistance depends upon the reference or starting temperature: it

TABLE 18.1

Approximate values of the temperature coefficient of resistance for various materials at $0 \,°C$

Material	Temperature coefficient of resistance α_0
Aluminium	42×10^{-4}
Carbon	-5×10^{-4}
Copper	43×10^{-4}
Lead	43×10^{-4}
Manganin	0
Mercury	7×10^{-4}
Nickel	56×10^{-4}
Tungsten	49×10^{-4}
Zinc	43×10^{-4}

EXAMPLE 18.1. A cable 50 m long has two copper conductors each of 6.7 mm^2 cross-sectional area, and carries a current of 30 A. Estimate the voltage drop in the cable at $25 \,°C$ if the resistivity of the copper at $0 \,°C$ is $17 \; \mu\Omega$ mm and the temperature coefficient of resistance at $0 \,°C$ is $0.004 \, 3/ \,°C$.

SOLUTION

The cable is 50 m long so that the total length of conductor is 2×50 m $= 100$ m $= 100 \, 000$ mm.

Resistance $R = \rho \dfrac{l}{A}$, where the resistivity is given at $0 \,°C$.

Hence at $0 \,°C$, $R_0 = \dfrac{17 \times 100 \, 000}{10^6 \times 6.7}$

$\qquad\qquad = 0.252 \; \Omega$

will be seen that for a given material α_{20}, α_{50}, etc., are not the same as α_0. For this reason, if a reference temperature θ is to be used, the value of α_θ must be known.

Let R_θ = resistance at 'reference' temperature θ °C,

$\quad R_2$ = resistance at some other temperature θ_2 °C,

$\quad \alpha_\theta$ = temperature coefficient of resistance from and at θ °C.

Substituting in eqn (1):

$$R_2 = R_\theta\{1+\alpha_\theta(\theta_2-\theta)\}. \qquad (5)$$

Then $\dfrac{R_2}{R_\theta} = 1+\alpha_\theta(\theta_2-\theta).$ $\qquad (6)$

But for eqn (4):

$$\frac{R_2}{R_\theta} = \frac{1+\alpha_0\,\theta_2}{1+\alpha_0\,\theta}. \qquad (7)$$

Equating (6) and (7):

$$1+\alpha_\theta\,(\theta_2-\theta) = \frac{1+\alpha_0\,\theta_2}{1+\alpha_0\,\theta},$$

$$\alpha_\theta\,(\theta_2-\theta) = \frac{1+\alpha_0\,\theta_2}{1+\alpha_0\,\theta}-1$$

$$= \frac{(1+\alpha_0\,\theta_2)-(1+\alpha_0\,\theta)}{1+\alpha_0\,\theta}$$

$$= \frac{\alpha_0(\theta_2-\theta)}{1+\alpha_0\,\theta},$$

$$\therefore\ \alpha_\theta = \frac{\alpha_0}{1+\alpha_0\,\theta}. \qquad (8)$$

EXAMPLE 18.3. If the resistance of a length of nickel wire is 53 Ω at 20 °C and 61 Ω at 50 °C, calculate the temperature coefficient of resistance of the nickel from and at 20 °C. Compare this value with that from and at 0 °C.

SOLUTION

Substituting the given values into

$$R_2 = R_\theta\{1+\alpha_\theta(\theta_2-\theta)\},$$

$$61 = 53\{1+\alpha_{20}(50-20)\},$$

$$\frac{61}{53} = 1+30\alpha_{20},$$

$$\alpha_{20} = \frac{1\cdot15-1}{30}$$

$$= 0\cdot005,$$

whence temperature coefficient of resistance from and at 20 °C = 0·005/°C.

From the relationship

$$\alpha_\theta = \frac{\alpha_0}{1+\alpha_0\,\theta},$$

$$\alpha_0 = \frac{\alpha_\theta}{1-\alpha_\theta\,\theta}$$

$$= \frac{0\cdot005}{1-(0\cdot005\times20)}$$

$$= \frac{0\cdot005}{0\cdot9}$$

$$= 0\cdot005\,6.$$

Temperature coefficient of resistance from and at 0 °C, = 0·005 6/°C.

The difference in values of 0·000 6 represents a change of 12 per cent.

SUMMARY

In general, the resistance of materials changes with increase in temperature as follows:

(1) conductors: increases;
(2) insulators: decreases;
(3) 'resistance alloys': practically constant.

The temperature coefficient of resistance of a material is the change in resistance per ohm per degree temperature rise.

The unit of α is /°C.

$R_\theta = R_0(1+\alpha_0\,\theta)$, where R_0 = resistance at 0 °C, R_θ = resistance at some other temperature θ °C, α_0 = temperature coefficient of resistance from and at 0 °C.

$\dfrac{R_1}{R_2} = \dfrac{1+\alpha_0\,\theta_1}{1+\alpha_0\,\theta_2}$, where R_1 = initial resistance at temperature θ_1 °C, R_2 = final resistance at temperature θ_2 °C.

$$R_2 = R_\theta\{1+\alpha_\theta(\theta_2-\theta)\},$$

and $\alpha_\theta = \dfrac{\alpha_0}{1+\alpha_0\,\theta}$, where R_θ = resistance at 'reference' temperature θ °C, R_2 = resistance at some other temperature θ_2 °C, α_θ = temperature coefficient of resistance from and at reference temperature θ °C.

EXERCISES 18

1. Name, in each case, *one* material whose electrical resistance (a) increases with a rise of temperature, (b) decreases with a rise of temperature.

Most relevant section: 18.1

U.L.C.I.

2. Write a few lines explaining the effect of temperature on the electrical resistance of metals. Define the term 'temperature coefficient of resistance'.

Most relevant section: 18.1.1

E.M.E.U.

3. How does the electrical resistance of most conductors vary as their temperature changes?

Most relevant section: 18.1

U.E.I.

4. What *four* factors determine the value of a resistor?

Most relevant sections: 17.9.3 and 18.1

U.E.I.

5. State and explain one application of a material which has a negligible temperature coefficient of resistance. The resistance of a coil of copper wire is 24 Ω at 20 °C. Determine its resistance at 70 °C if the temperature coefficient of resistance referred to 0 °C is 0·004 26/°C.

Most relevant sections: 18.1 and 18.1.1

Y.C.F.E.

6. What is understood by the term 'temperature coefficient of resistance' of a conductive material? The resistance of a coil of copper wire at 30 °C is 43·7 Ω and at 62 °C is 49 Ω. Calculate the temperature coefficient of copper referred to 0 °C and state its units.

Most relevant section: 18.1.1

Y.C.F.E.

7. A coil takes a current 1·5 A from a supply of 240 V when switched on at room temperature (15 °C). After being connected to the supply for a time the current is noticed to have fallen to 1·4 A. Calculate the temperature of the coil.

Temperature coefficient of resistance

$$= 0·004\ 3/°C.$$

Most relevant sections: 17.5 and 18.1.1

U.C.L.I.

8. The following results were obtained in an experiment to find the temperature coefficient of resistance of a length of nickel wire.

Temperature (°C)	10	20	30	40	50
Resistance (Ω)	50	53	56	59	62

Plot a graph of resistance against temperature and use this graph to determine the temperature coefficient of resistance from 0 °C.

Most relevant section: 18.1.1

U.E.I

9. A constant voltage of 240 V is applied to the field windings of a motor. At 15 °C the current flowing through the windings is 2·0 A. After the motor had been running for some time the field current was found to be 1·9 A. Calculate the temperature of the field windings given that the temperature coefficient of resistance for the coil windings is 0·004 2/°C.

Most relevant sections: 17.5 and 18.1.1

E.M.E.U.

10. What is meant by 'temperature coefficient of resistance'?
The resistance of a length of platinum wire is 5·2 Ω at 0 °C and 7·28 Ω at 100 °C. If the variation of resistance between the two temperatures is uniform, determine:
(a) the temperature coefficient of resistance,
(b) the temperature of the wire when its resistance is 6·5 Ω.

Most relevant section: 18.1.1

U.E.I.

11. (a) Explain the statement '*the temperature coefficient of resistance of carbon from* 0 °C *is* $-5 \times 10^{-4}/°C$'. Sketch a graph of *resistance against temperature* to illustrate your answer.
(b) The resistance of a coil of nickel wire is 53 Ω at 20 °C. Determine its resistance at 50 °C if the temperature coefficient of resistance from 0 °C is $5·6 \times 10^{-3}$ /°C.

Most relevant section: 18.1.1

U.E.I.

12. (a) What four factors determine the electrical resistance of a piece of material.
(b) The resistance of a certain tungsten filament lamp is 1000 Ω at its working temperature of 2000 °C. Estimate its resistance when at a room temperature of 15 °C. Take the temperature coefficient of resistance of tungsten as 0·005/°C at and from 0 °C.

Most relevant sections: 17.9 and 18.1.1

U.E.I.

13. (a) How does the electrical resistance of most conductors vary with:

  (i) increase in length,
  (ii) increase in cross-sectional area, and
  (iii) increase in temperature.

(b) A solenoid has a resistance of 480 Ω at 20 °C. Calculate its resistance at 87·5 °C if the temperature coefficient of resistance of the wire is 0·004/°C at and from 0 °C.

Most relevant sections: 17.9 and 18.1.1

U.E.I.

14. (a) If the length of a conductor is doubled, while its cross-section remains unaltered, how does the value of its resistance vary?

(b) A cable 100 m long has a cross-sectional area of 8·5 mm² and carries a direct current of 30 A. Calculate the voltage drop along the cable at 20 °C if the resistivity of the copper at 0 °C is 17 μΩ mm and the temperature coefficient of resistance of copper at 0 °C is 0·0043/°C.

Most relevant sections: 17.9, 17.5, and 18.1.1

U.E.I.

15. (a) Define the terms 'resistivity' and 'temperature coefficient of resistance' of a material.

(b) If the resistance of a length of nickel wire is 53 Ω at 20 °C and 61 Ω at 50 °C, estimate the temperature coefficient of resistance of the nickel at 0 °C.

Most relevant sections: 17.9 and 18.1.1

U.E.I.

ANSWERS TO EXERCISES 18

5. 28·7 Ω. **6.** 0·004 26/°C. **7.** 31·8 °C. **8.** 0·006 37/°C.
9. 28·7 °C. **10.** (a) 0·004/°C, (b) 62·5 °C. **11.** (b) 61 Ω.
12. (b) 97·7 Ω. **13.** (b) 600 Ω. **14.** (b) 6·516 V. **15.** (b) 0·005 595/°C.

19 Laws of electrolysis

19.1. ELECTROLYSIS

The flow of a current through a solid conductor, such as copper wire, is accompanied by heating and magnetic effects in the conductor but no chemical changes occur.

If, however, a current is passed through a liquid conductor a chemical action takes place and the liquid is dissociated into its constituent parts.

This decomposition of the liquid conductor is termed electrolysis.

19.1.1. Definitions. *Electrolysis* is the decomposition of a solution or molten compound by the passage through it of an electric current.

An *electrolyte* is a solution or molten compound which will conduct electricity: chemical changes always occur in the solution when an electric current is passed.

Electrodes are metal or carbon plates by which the current enters or leaves an electrolyte.

The *anode* is the positive electrode by which the current enters the electrolyte.

The *cathode* is the negative electrode by which the current leaves the electrolyte.

Many liquids are conductors of electricity, but as in the case of solids, some liquids are better conductors than others. The best liquid conductors are solutions of simple salts, e.g. sodium chloride; acids, e.g. sulphuric acid or alkalis, e.g. caustic soda.

Pure liquids such as distilled water and alcohol do not conduct electricity.

19.1.2. Chemical effects at the electrodes. Two main actions take place during electrolysis:
1. Metals and hydrogen are liberated at the cathode.
2. Non-metals and acid radicals are liberated at the anode.

19.2. THE IONIC THEORY

All electrolytes in solution are thought to split up into two parts, each carrying an electric charge. These charged parts are called *ions*.

The ions of metals and hydrogen are positively charged due to a lack of electrons: the ions of non-metals are negatively charged.

When a potential difference is applied to electrodes dipping into an electrolyte the current is carried through the solution by a drift of ions towards the electrodes. As unlike charges attract, the negative ions move towards the anode and the positive ions move to the cathode.

After reaching the electrodes the ions give up their charges and are liberated as ordinary atoms.

Note: Since metallic ions carry positive charges they are attracted to the negative electrode. At this electrode (cathode) the ions give up their electrical charge and a metal deposit is found on the cathode.

19.2.1. Electrolysis of copper sulphate. The copper sulphate solution (electrolyte) splits up into positive copper ions and negative sulphate ions. When a potential difference is applied to the electrodes (copper), the positively charged copper ions are attracted to the cathode where they lose their charge and are deposited forming a layer of copper (see Fig. 19.1).

The negatively charged sulphate ions move towards the anode where they are discharged and combine with the copper anode to form copper sulphate. In this way the copper-sulphate electrolyte is replenished but the anode is slowly eaten away.

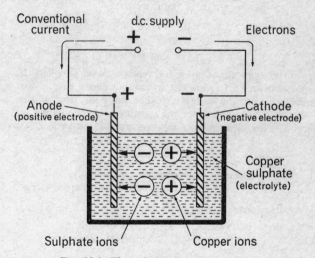

FIG. 19.1. Electrolysis of copper sulphate

19.3. APPLICATIONS OF ELECTROLYSIS

19.3.1. Electro-plating. In the electro-plating industry electrolytic methods are used to coat one metal with a thin layer of another. This process is known as electro-plating. The reasons for electro-plating objects are:

(a) to protect certain metals from corrosion,
(b) to improve appearance.

An example of (a) is the process of plating steel with nickel and chromium.

The surface provided is both protective and attractive and among others, has wide application in the motor car industry.

19.3.2. Metal refining. The electrolytic refining of copper is carried out on a large scale industrially and the pure copper obtained is used extensively in the electrical industry.

In this process thin sheets of pure copper form the cathode and the impure copper acts as the anode: the electrolyte is acidified copper sulphate. During electrolysis the impure copper leaves the anode and is deposited as pure copper on the cathode (see electrolysis of copper sulphate).

Similar methods are employed to refine lead and nickel.

19.4. FARADAY'S LAWS OF ELECTRO-LYSIS

An English physicist *Michael Faraday* (1830) discovered that the mass of a substance liberated at an electrode during electrolysis is proportional to:

(1) the magnitude of the *current* passing;
(2) the *duration* of the current (in seconds).

If m is the mass of the substance liberated,

  t the time for which the current passes, and
  I the magnitude of the current,

then Faraday's laws expressed in symbols become

$$m \propto It,$$
$$\therefore m = \text{constant} \times It.$$

This constant is given the symbol z and is called the *electrochemical equivalent of a substance (e.c.e.)*. For convenience the term electrochemical equivalent will be abbreviated to e.c.e. for the remainder of the chapter.

The above equation may therefore be written as

$$m = zIt.$$

In words,

**mass of substance liberated at an electrode
= e.c.e. of substance × current × time.**

19.4.1. Units. In the previous chapter it was stated that

$$\begin{array}{c}\text{quantity of electricity} \\ (Q \text{ coulombs})\end{array} = \begin{array}{c}\text{current} \\ (I \text{ amperes})\end{array} \times \begin{array}{c}\text{time} \\ (t \text{ seconds})\end{array}$$

or $$Q(\text{C}) = I(\text{A}) \times t(\text{s}).$$

The equation $m = zIt$ may therefore be written as

$$m = zQ \qquad (Q = It).$$

Rearranging,

$$z = \frac{m}{Q}.$$

If mass m is expressed in grams (g),
  current I is expressed in amperes (A), and
  time t is expressed in seconds (s),

then $$z = \frac{m(\text{g})}{Q(\text{C})}$$

or z is expressed in grams/coulomb (g/C),

$$\therefore m(\text{g}) = z(\text{g/C}) \times I(\text{A}) \times t(\text{s}).$$

The electrochemical equivalent of a substance is defined as the mass of any element deposited or liberated by 1 coulomb of electricity.

Table 19.1 gives the electrochemical equivalents of a number of common elements.

TABLE 19.1

Element	Electrochemical equivalent (z) (g/C)
Cadmium	0·000 582
Chromium	0·000 089 8
Copper	0·000 329
Hydrogen	0·000 010 45
Nickel	0·000 304
Oxygen	0·000 082 9
Platinum	0·000 506
Silver	0·001 118
Tin	0·000 308
Zinc	0·000 339

EXAMPLE 19.1. A current of 15 A flows through an electrolyte of copper sulphate for 10 min.

Calculate the mass of copper deposited on the cathode if the e.c.e. of copper is 0·000 33 g/C.

SOLUTION

$$z = 0.000\ 33 \text{ g/C},$$
$$I = 15 \text{ A},$$
$$t = 10 \times 60 \text{ s} = 600 \text{ s}.$$

Required to find m,

  using $m = zIt$,

  then $m = 0.000\ 33 \times 15 \times 600$ g

  $$= 2.97 \text{ g}.$$

Mass of copper deposited on the cathode = 2·97 g.

EXAMPLE 19.2. In a certain electrolytic copper refining process a current of 5000 A was used. Calculate
(a) the time required to refine 1 kg of copper,
(b) the quantity of electricity used expressed (i) in ampere-hours (ii) in coulombs.
The e.c.e. of copper = 0·000 33 g/C.

SOLUTION

(a)
$$m = 1 \text{ (kg)} = 1000 \text{ (g)},$$
$$z = 0.000\ 33,$$
$$I = 5000.$$

Required to find t,

using $m = zIt$ and rearranging for t,

$$\text{time } t = \frac{m}{zI}$$

$$= \frac{1000}{0.000\ 33 \times 5000}$$

$$= \frac{1000}{1.65} = 606 \text{ s}.$$

$$\therefore t = \frac{606}{60} \text{ (min)} = 10 \text{ min } 6 \text{ s}.$$

Time required to refine 1000 g (1 kg) copper

$$= 10 \text{ min } 6 \text{ s } (10.1 \text{ min}).$$

(b) (i) Number of ampere-hours required

$$= \text{current (A)} \times \text{time (h)}$$

$$= 5000 \text{ A} \times \frac{10.1}{60} \text{ h}$$

$$= \frac{5050}{6} = 841.7 \text{ A h}.$$

Quantity of electricity required $= 841.7$ A h.

(ii) Number of coulombs required

$$= \text{current (A)} \times \text{time (s)}$$

$$= 5000 \text{ A} \times 606 \text{ s}$$

$$= 3\ 030\ 000 \text{ C}.$$

Expressed in coulombs the quantity

of electricity required $= 3\ 030\ 000$ C.

Note: In this case it can be seen that where large currents and times are involved it is more convenient to express quantity of electricity in terms of the ampere-hour.

EXAMPLE 19.3. A metal plate having a surface area of 1000 cm² is to be coated by an electrolytic process with nickel to a depth of 0·1 mm. Calculate the quantity of electricity required in ampere-hours.
Density of nickel $= 8.7$ g/cm³,
electrochemical equivalent of nickel $= 0.000\ 304$ g/C.

SOLUTION

In this question the mass of nickel required is not given directly but may be calculated as follows:

$$\text{mass} = \text{volume} \times \text{density}$$

$$= 1000 \times \frac{0.1}{10} \times 8.7 \qquad \left(0.1 \text{ mm} = \frac{0.1}{10} \text{ cm}\right),$$

$$\therefore m = 87.$$

Using $m = zQ$ and rearranging for Q,

$$Q = \frac{m}{z} = \frac{87}{0.000\ 304}$$

$$= 286\ 200 \text{ (C)}$$

$$= \frac{286\ 200}{60 \times 60} = 79.5 \text{ (A h)}.$$

$\therefore$ Quantity of electricity required $= 79.5$ A h.

SUMMARY

Electrolysis is the decomposition of a solution or molten compound by the passage through it of an electric current.

An electrolyte is a liquid which conducts electricity and is decomposed in the process.

The anode is the positive electrode by which the current enters the electrolyte.

The cathode is the negative electrode by which the current leaves the electrolyte.

Applications of electrolysis include metal refining and electroplating.

Faraday's law of electrolysis may be expressed as follows:

The mass of a substance liberated at an electrode during electrolysis is proportional to the quantity of electricity passed through the electrolyte.

$$\text{or } m \propto Q,$$

alternatively $m \propto It$ (since $Q = It$).

From Faraday's discoveries the fundamental equation,

$$m = zIt \text{ is obtained}$$

where z is the electrochemical equivalent of the substance.

m is normally expressed in grams,
z is normally expressed in grams/coulomb,
I is expressed in amperes,
t is in seconds.

The electrochemical equivalent of a substance is the mass of the substance liberated by one coulomb of electricity.

EXERCISES 19

1. If a deposit of 0·18 g of metal forms on a cathode in 10 min and a current of 1 A is flowing, what is the electrochemical equivalent of this metal?

Most relevant section: 19.4

U.E.I.

2. Define the 'coulomb'.
   In an experiment with a copper voltameter the initial mass of the cathode was 40 g. A current of 5 A was then passed through the voltameter for a period of 1 h. If the electrochemical equivalent of copper = 0·000 33 g/C, determine:
   (a) the number of coulombs of electricity which passed through the apparatus;
   (b) the new mass of the cathode.

Most relevant sections: **17.1.1** and **19.4**

U.C.L.I.

3. A copper voltameter, an accumulator, and an ammeter are connected in series. In an hour 2 g of copper are deposited. Determine the reading that the ammeter should show.
   (Electrochemical equivalent of copper = 0·000 33 g/C.)
   By means of a diagram show the way in which the apparatus was connected and indicate:
   (a) the direction of conventional current flow;
   (b) the polarity of the accumulator;
   (c) the anode of the voltameter.

Most relevant sections: **19.4** and **19.2.1**

N.C.T.E.C.

4. A copper voltameter connected in series with an ammeter was used as the means of checking the calibration of the ammeter. During the test the ammeter reading was 2 A and in 15 min the mass of copper deposited was 0·625 g. Take the electrochemical equivalent of copper as 0·000 33 g/C and calculate:
   (a) the true current;
   (b) the error in the ammeter reading;
   (c) the quantity of electricity which flowed during the test.

Most relevant section: **19.4**

U.E.I.

5. (a) State Faraday's first law of electrolysis.
   (b) If the electrochemical equivalent of silver is 0·001 118 g/C, calculate the mass of silver deposited on the cathode of a silver-plating vat in 6 h while a current of 2·4 A is flowing.

Most relevant section: **19.4**

U.E.I.

6. What is a coulomb of electricity?
   A steady current of 3 A is passed for 20 min through a silver nitrate voltameter. Determine the mass of silver deposited on the cathode in this time and specify the quantity of electricity used, if the electrochemical equivalent of silver is 0·001 118 g/C.

Most relevant sections: **17.1.1** and **19.4**

W.J.E.C.

7. Calculate the quantity of zinc used when a Leclanché cell supplies a current of 0·05 A for 1 h.
   ($z = 0\cdot339$ mg/C.)

Most relevant section: **19.4**

N.C.T.E.C.

8. How long will it take to deposit 25 g of silver in an electro-plating process using a current of 1·5 A? The electrochemical equivalent of silver is 1·12 mg/C.

Most relevant section: **19.4**

C.G.L.I.

9. Determine the mass of copper deposited on the cathode of a copper-sulphate voltameter when a current of 1·2 A is passed for 20 min. (Electrochemical equivalent of copper is 0·000 33 g/C.)

Most relevant section: **19.4**

Y.C.F.E.

10. Explain precisely what happens when water is electrolysed.
    A metal plate having a total surface area of 90 cm² is to be copper plated to a depth of 0·04 mm. Calculate the time required to deposit this thickness if a plating current of 2 A is used.
    Density of copper = 8·93 g/cm³.
    Electrochemical equivalent of copper = 0·000 329 4 g/C.

Most relevant sections: **1.2.1** and **19.4**

E.M.E.U.

ANSWERS TO EXERCISES 19

1. 0·000 3 g/C. **2.** (a) 18 000 C, (b) 45·94 g. **3.** 1·68 A. **4.** (a) 2·104 A, (b) 0·104 A. (c) 1893·6 C. **5.** 58 g. **6.** 4·02 g, 3600 C. **7.** 61·02 mg. **8.** 4·14 h. **9.** 0·455 g. **10.** 1·35 h.

20 Primary and secondary cells

20.1. CELLS

One source of *electromotive force* (*e.m.f.*) is a *cell* or *battery*. In a cell electricity is generated by the conversion of chemical energy into electrical energy. Cells are divided into two main groups known as primary and secondary cells.

20.2. PRIMARY CELLS

The word 'primary' is used to describe a cell in which electricity is produced at the expense of the chemicals from which the cell is constructed and *once these chemicals have been used up, no more electricity can be obtained from the cell.*

The essential parts of a cell are two electrodes which are dissimilar metals, or carbon and a metal, and an electrolyte.

20.2.1. Principles of the voltaic cell. The *voltaic cell*, devised by an Italian scientist *Volta* (1745–1827) consists of:

(1) a plate of zinc forming the negative electrode (cathode);
(2) a plate of copper as the positive electrode (anode);
(3) dilute sulphuric acid as the electrolyte.

The arrangement of the cell is shown in Fig. 20.1. The two electrodes are immersed in the dilute sulphuric acid.

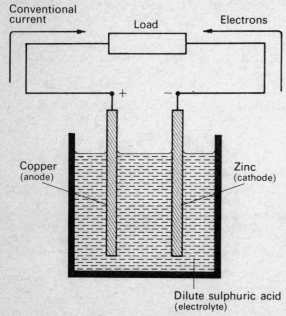

Fig. 20.1. Voltaic cell

When the electrodes are connected outside the cell, a current flows from the copper through the external circuit to the zinc and from the zinc to the copper through the cell.

The e.m.f. obtained from a cell of this type is approximately 1 V.

DEFECTS OF THE SIMPLE CELL

Local action is a term used to describe the chemical reaction which takes place between the impure zinc plate forming the negative electrode and the sulphuric acid, resulting in the zinc being dissolved in the acid.

The reaction between the impure zinc and dilute sulphuric acid takes place in the cell irrespective of whether a current is taken from the cell or not.

Local action, which is obviously wasteful, may be eliminated by coating the zinc plate with mercury.

Note: Pure zinc does not dissolve in dilute sulphuric acid but its comparatively high cost prevents its use in the simple cell.

The current from a simple cell rapidly falls off due to the formation of a layer of hydrogen gas on the positive copper electrode.

The formation of this layer of hydrogen on the positive electrode is known as *polarization*.

Polarization has the effect of reducing the e.m.f. of the cell and increasing its internal resistance.

The removal of the hydrogen gas layer from the copper electrode ('depolarization') may be achieved by either:

(1) a mechanical method which involves rubbing or brushing the electrode, or
(2) a chemical method which employs a material rich in oxygen to oxidize the hydrogen to form water.

20.2.2. Leclanché cells.

WET TYPE

This cell (Fig. 20.2), invented by the French scientist *Georges Leclanché*, consists of a square-shaped glass vessel which contains:

(1) a rod of zinc coated with mercury forming the negative electrode;
(2) a rod of carbon as the positive electrode surrounded by a mixture of powdered carbon and manganese dioxide contained in a porous pot;
(3) an electrolyte of strong ammonium chloride.

The amalgam (mercury coating) on the zinc rod eliminates local action.

The oxygen from the manganese dioxide combines

with the hydrogen liberated at the positive electrode to form water. Hence the hydrogen-gas layer is continually removed from the carbon rod, so preventing polarization. The powdered carbon provides a good conducting path between the ammonium-chloride electrolyte and the positive carbon electrode.

When a current is taken from the cell polarization occurs at a faster rate than depolarization. The e.m.f. of the cell therefore begins to fall off. However, removal of the hydrogen gas layer continues after the circuit is broken. Consequently the cell is more suitable for intermittent work.

The e.m.f. of the Leclanché cell is about 1·5 V.

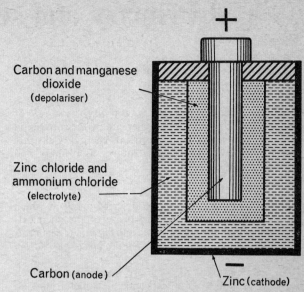

Fig. 20.3. Leclanché (dry) cell

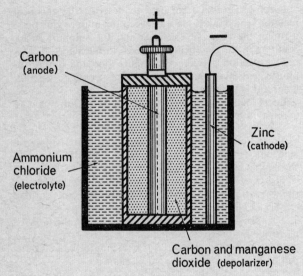

Fig. 20.2. Leclanché (wet) cell

DRY TYPE

This dry type of Leclanché cell (Fig. 20.3) is the most widely used of primary cells and consists of:

(1) a zinc container forming the negative electrode;
(2) a carbon rod as the positive electrode surrounded by crushed carbon and manganese dioxide—the depolarizer;
(3) a moist paste of plaster of Paris, zinc chloride, and ammonium chloride—the electrolyte.

The zinc container is surrounded by cardboard for insulation and decorative purposes, and to prevent drying, the top is filled with sand and sealed with pitch.

The chemical action is similar to that of the wet cell. When new, the e.m.f. of the cell is about 1·4 V, but it quickly falls to ˙1·4 V and then remains fairly constant until the cell is near exhaustion.

This type of cell is extremely useful because of its portability and compact form and is used in electric torches, transistor radio sets, etc.

20.2.3. Mercury cell. This type of primary cell was developed to meet the demands of miniaturization in applications such as missiles and satelites, photographic equipment, and medical electronics. The materials used in the mercury cell provide it with many times the energy output of ordinary cells of comparable size. This has enabled mercury cells to be designed to provide the same capacity as conventional cells, but to be only one-third of the size. Mercury cells are produced in cylindrical and flat pellet structure. Electrochemically both forms are identical and differ only in container design and internal arrangement. The basic construction of the cylindrical form is illustrated in Fig. 20.4.

The negative electrode is formed from high-purity

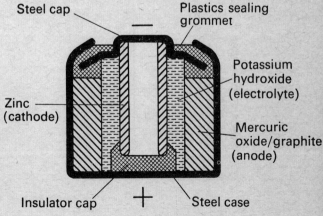

Fig. 20.4. Mercury cell

amalgamated zinc powder pressed into pellets. The positive electrode is mercuric oxide with graphite and separated from the cathode by an ion-permeable barrier which prevents migration of active material within the cell. The electrolyte is a solution of potassium hydroxide, an alkali, whose ions act as carriers for the chemical action of the cell, but is not consumed. In operation this combination produces a neutral film of mercury which does not inhibit current flow. Thus the terminal voltage remains steady on discharge and recuperation periods are unnecessary. To ensure maximum realization of the energy available, the inner cell top is plated to provide an internal surface with which the zinc electrode is electrochemically compatible. Cell containers are nickel-plated steel to resist corrosion and offer the greatest passivity to the electrolyte.

The nominal cell voltage is 1·35–1·4 V. When the cell is at work the cell voltage forces electrons to flow through the external circuit in accordance with Ohm's law. Loss of electrons through the external circuit disturbs the cathode equilibrium. Hydroxyl ions from the electrolyte discharge at the cathode to restore the cathode charge, forming zinc oxide and water. Gain of electrons upsets the anode equilibrium. Hydrogen ions from the electrolyte discharge at the anode, forming mercury and water. There is no net change of water content in the electrolyte.

Some examples of the advantages of the mercury cell are given below.

The ratio of energy to volume and to weight is high.

There is negligible local action and under normal operating conditions there is no polarization. Cells can be stored for periods up to 2½ years in dry conditions and at temperatures around 20 °C (293 K) without any appreciable loss of capacity.

Being completely enclosed in sealed nickel-plated steel containers the cell can withstand severe vibration, shock and acceleration forces without voltage fluctuation.

High vacuum, or pressures up to 35 MPa, have no detectable effect on the cell.

20.3. SECONDARY CELLS

The function of the primary cell discussed in the previous section is to convert chemical energy into electrical energy at the expense of the electrodes and electrolyte forming the cell.

In a 'secondary' cell, or 'accumulator', electrical energy initially supplied from an external source is *converted and stored* in the cell as chemical energy. This *conversion of energy is reversible* and when required the chemical energy so stored can be released as a direct electric current.

When an electric current is being taken from the cell it is said to be 'discharging'. The reverse process of supplying electrical energy to the cell is called 'charging'.

20.3.1. The lead–acid accumulator. The most widely used of the secondary cells is the lead–acid type. Two or more cells joined together is termed a battery and in this form has many commercial uses.

Today, however, motor car and vehicle batteries account for about 80 per cent of the lead used in the battery industry. The typical car battery, for example, consists of six lead–acid cells joined together.

CONSTRUCTION

A lead–acid cell consists of:

(1) a lead grid, or framework, covered with lead peroxide which forms the positive plate;
(2) a lead grid containing spongy lead which forms the negative plate; and
(3) an electrolyte of dilute sulphuric acid.

PRINCIPLES OF OPERATION

The principles of operation of lead–acid accumulators are outlined in Figs 20.5 (a)–(d).

DATA ON LEAD-ACID ACCUMULATOR

A fully charged lead–acid cell gives an e.m.f. of approximately 2·2 V but when in use this value falls rapidly to about 2 V.

The cell is said to be discharged after the e.m.f. has fallen to 1·8 V and requires recharging before further use. Fully charged the relative density of the sulphuric acid should be about 1·25, when discharged this falls to about 1·1.

20.3.2. Alkaline cell. The alkaline cell is another extensively used secondary cell. Fig. 20.6 shows the general construction of one type of alkaline cell known as the nickel–cadmium cell.

The positive plate contains mainly nickel hydroxide and the negative plate contains a mixture of cadmium and iron. The electrolyte is a solution of potassium hydroxide, an alkali.

The electrolyte and electrodes are contained in a welded steel casing.

This cell gives an e.m.f. of 1·2 V.

CHARACTERISTICS OF ALKALINE CELLS

(1) Very robust both mechanically and electrically.
(2) Unlimited rates of charge and discharge. Short circuits will not damage the battery.
(3) Very little deterioration even when left discharged for long periods.
(4) Long life when correctly maintained.

The cell possesses certain disadvantages which include:

(a) high initial cost compared with lead–acid cell;
(b) the internal resistance of the cell is higher than the lead acid cell;
(c) the discharge voltage is only 1·2 V per cell compared with 2 V for the lead-acid cell.

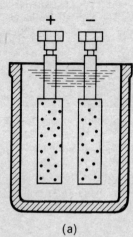

(a)

Discharged. Positive plate: active material mainly lead
  sulphate.
Negative plate: active material mainly lead sulphate.
Electrolyte: minimum acid content and maximum water.
Minimum relative density.

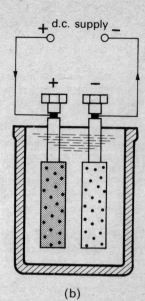

(b)

Cell charging. Positive plate: lead sulphate being
  converted to lead peroxide.
Negative plate: lead sulphate being converted by the
  current to porous metallic lead.
Electrolyte: sulphuric acid content increases. Relative
  density increases.

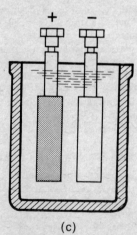

(c)

Cell charged. Positive plate: lead peroxide. Negative
  plate: porous metallic lead.
Electrolyte: maximum acid content. Maximum relative
  density.

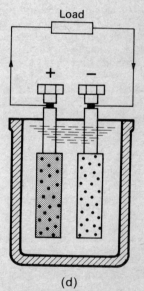

(d)

Cell discharging. Positive plate: lead peroxide being
  converted to lead sulphate. Water formed.
Negative plate: porous metallic lead being converted to
  lead sulphate. Water formed.
Electrolyte: acid content falls due to formation of water.
Relative density falls.

FIG. 20.5. Principles of operation of lead–acid cells

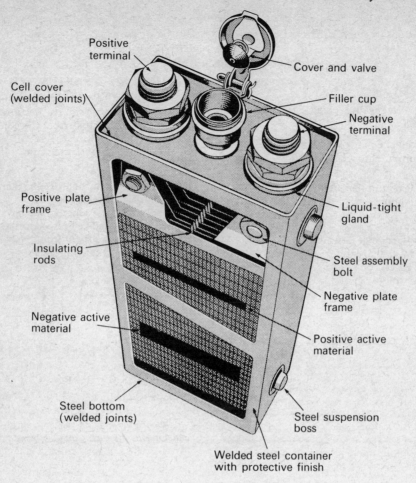

Fig. 20.6. Nickel–cadmium cell

These cells have a wide range of application from hand-lamps to automatic train-control systems, etc.

20.4. INTERNAL RESISTANCE OF A CELL

Every cell offers some resistance to the passage of an electric current. The amount of resistance offered depends upon the nature of the positive and negative electrodes or plates and the type of electrolyte used.

This resistance offered by a cell is termed its 'internal resistance'.

20.4.1. Voltage drop due to internal resistance. As previously stated the e.m.f. E (volts) of a cell is the total voltage generated by a cell on open circuit.

Now let r be the internal resistance of the cell and I the current flowing when the cell is connected to an external circuit.

The voltage drop due to the cell resistance with current I flowing

$$= Ir \text{ (Ohm's law: p.d. } = Ir).$$

Hence, p.d. of cell

$$= \text{e.m.f. of cell} - (\text{current} \times \text{internal resistance}),$$

or, in symbols,

$$V = E - Ir.$$

Note:

$$V = E - Ir = IR,$$

where R is the resistance of the external circuit.

20.5. CIRCUIT SYMBOL FOR A CELL

Fig. 20.7 shows the symbols used to represent an 'ideal' cell and a 'real' cell. The ideal cell is one which is assumed to have no internal resistance.

The thin long line represents the positive pole and the

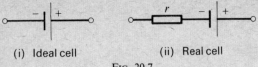

(i) Ideal cell   (ii) Real cell

Fig. 20.7

short thicker line the negative pole: r is the internal resistance of the cell.

EXAMPLE 20.1. A cell of e.m.f. 2 V and internal resistance $0.01\ \Omega$ drives a current through a circuit of resistance $0.09\ \Omega$.

Calculate (a) the current in the circuit, and
            (b) the terminal p.d. of the cell.

SOLUTION

Fig. 20.8 shows the circuit in which e.m.f. $(E) = 2$ V, internal resistance $(r) = 0.01\ \Omega$, and resistance of external circuit $(R) = 0.09\ \Omega$.

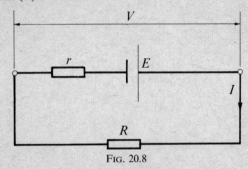

Fig. 20.8

(a) Total circuit resistance (including cell resistance);
$$R+r = 0.09+0.01.$$
Total circuit resistance $= 0.1\ \Omega$.
$$\text{Using } V = E - Ir = IR,$$
$$\text{then } E = Ir+IR,$$
$$E = I(r+R),$$
$$\text{and } I = \frac{E}{R+r} = \frac{2}{0.1} = 20.$$
Current in the circuit $= 20$ A.

(b) Terminal p.d. of cell may be found by using either
$$\text{(a) } V = E - Ir \text{ or (b) } V = IR.$$
(a) Using $V = E - Ir$,
$$V = 2 - 20 \times 0.01$$
$$= 2 - 0.2$$
$$= 1.8.$$
Terminal p.d. $= 1.8$ V.
(b) Using $V = IR$,
$$V = 20 \times 0.09$$
$$= 1.8.$$
Terminal p.d. $= 1.8$ V.

Note: The e.m.f. of the cell is 2 V, but the p.d. between the terminals of the cell is only 1·8 V. This voltage difference is due to the internal resistance of the cell.

20.6. CELLS IN SERIES

Cells are said to be in *series* when the *positive plate of one cell is connected to the negative plate of another* as in Fig. 20.9. A number of cells connected in this way is termed a battery.

Cells arranged in series have the following characteristics:

(1) The battery e.m.f. equals the sum of the e.m.f.s of the individual cells.
(2) The internal resistance of the battery equals the sum of the individual cell resistances.
(3) The maximum current from the battery is the same as that from one individual cell.

Fig. 20.9. Cells in series

20.7. CELLS IN PARALLEL

A number of cells are said to be connected in *parallel* when all the *positive terminals are connected to one common terminal and all the negative terminals are connected to another common terminal* as in Fig. 20.10. Characteristics of cells in parallel are:

(1) The e.m.f. of several cells in parallel equals the e.m.f. of one individual cell.
(2) The total current equals the sum of the individual currents.
(3) The reciprocal of the total internal resistance equals the sum of the reciprocals of each internal resistance.

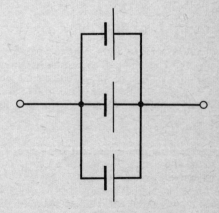

Fig. 20.10. Cells in parallel

EXAMPLE 20.2. An electric battery has an e.m.f. of 11 V and an internal resistance of 0·5 Ω.

A 2-Ω resistor A is placed in series with a 3-Ω resistor B across the battery terminals.

(a) What is the current taken from the battery?
(b) What is the potential difference across the battery terminals?
(c) What is the potential difference across each resistor?

U.E.I.

SOLUTION

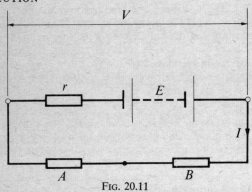

Fig. 20.11

Fig. 20.11 shows the circuit in which $E = 11$V, $r = 0.5\ \Omega$. For the external resistance, $R = 2\ \Omega + 3\ \Omega = 5\ \Omega$

(a) Using $E - Ir = IR$,
$$E = I(R+r),$$
$$11 = I(5+0.5),$$
$$11 = 5.5\ I,$$
and $I = \dfrac{11}{5.5} = 2$ A.

Current taken from the battery = 2 A.

(b) To find p.d. across battery terminals,
$$V = E - Ir$$
$$= 11 - 2 \times 0.5$$
$$= 11 - 1.$$
$$\therefore V = 10 \text{ volts.}$$
P.D. across battery terminals = 10 V.

(c) Let V_A be the p.d. across 2 Ω resistor, then
$$V_A = IR_A, \text{ where } R_A = 2\ \Omega,$$
$$\text{or } V_A = 2 \times 2 = 4 \text{ volts.}$$
P.D. across 2 Ω resistor = 4 V.
Let V_B be the p.d. across 3 Ω resistor, then
$$V_B = IR_B, \text{ where } R_B = 3\ \Omega.$$
$$\text{or } V_B = 2 \times 3 = 6 \text{ volts.}$$
P.D. across 3 Ω resistor = 6 V.

Note: The p.d. across the battery terminals equals 10 V = 6 V+4 V.

EXAMPLE 20.3. A battery consists of five cells in series, the e.m.f. of each cell being 1·4 V. If the battery supplies a current of 0·08 A to a circuit containing 2 resistors of 60 Ω and 25 Ω joined in series, calculate

(a) the internal resistance of each cell, and
(b) the terminal voltage of the battery.

SOLUTION

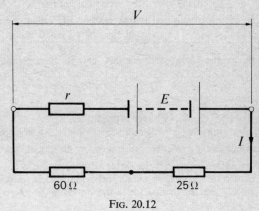

FIG. 20.12

Fig. 20.12 shows the circuit.

(a) The e.m.f. = 5×1.4 V = 7 V (cells in series). Resistance of external circuit (R) = 60 Ω+25 Ω (resistances in series).
$$\therefore R = 85\ \Omega.$$
$$I = 0.08.$$
$$\text{Current} = 0.08 \text{ A.}$$
Let the internal resistance of the battery (5 cells) = r.
Using $E - Ir = IR$,
$$7 - 0.08\ r = 0.08 \times 85,$$
$$7 = 0.08\ r + 0.08 \times 85,$$
$$7 = 0.08\ (r + 85),$$
$$\frac{7}{0.08} = r + 85,$$
$$87.5 = r + 85,$$
$$\therefore r = 87.5 - 85 = 2.5\ \Omega.$$

The resistance of 2·5 Ω represents the combined resistance of five cells in series.

Internal resistance of 1 cell = $\dfrac{2.5}{5}\ \Omega = 0.5\ \Omega$.

The internal resistance of each cell = 0·5 Ω.

(b) Terminal voltage of battery,
$$V = E - Ir,$$
$$\therefore V = 7 - 0.08 \times 2.5$$
$$= 7 - 0.2$$
$$= 6.8 \text{ volts.}$$
The terminal voltage of the battery = 6·8 V.

SUMMARY

In a cell electricity is generated by the conversion of chemical energy into electrical energy.

Cells are divided into two classes, primary cells and secondary cells.

Primary cells

$$\text{Chemical energy} \to \text{electrical energy}$$

The action of this cell is irreversible. Electrical energy is generated at the expense of the chemicals forming the cell.

The most widely used primary cell is the Leclanché cell.

Secondary cells

$$\text{Charge stored}$$
Cell charging   in the cell   Cell discharging
$$\text{Electrical energy} \rightleftharpoons \text{chemical energy} \rightleftharpoons \text{electrical energy}$$

This action is reversible. After discharge the cell can be restored to its original chemical condition by passing an electric current through it in the opposite direction to that of discharge.

The most widely used of the secondary cells are the lead–acid and alkaline cells, e.g. nickel–cadmium

Internal resistance of a cell

P.D. of a cell = e.m.f. of cell − $\left(\begin{array}{c}\text{current} \times \text{internal re-}\\ \text{sistance of cell}\end{array}\right)$.

In symbols, $V = E - Ir$, also $V = IR$,
where R is the resistance of the external circuit.

Cells in series

When the positive plate of one cell is connected to the negative plate of another the cells are 'in series'.

For cells in series:

(1) e.m.f. of battery equals sum of individual e.m.f.s,
(2) internal resistance of the battery equals sum of internal resistances; and
(3) the maximum current is the same as that for one cell.

Cells in parallel

Cells are 'in parallel' when the positive plates are connected to one common terminal and the negative plates are connected to another common terminal.

For cells in parallel:

(1) the e.m.f. of a number of cells in parallel is equal to the e.m.f. of one cell;
(2) the total current equals the sum of the individual currents; and
(3) the reciprocal of the combined internal resistance equals the sum of the reciprocals of each internal resistance.

EXERCISES 20

1. What is the basic difference between primary and secondary cells?

Most relevant sections: 20.2 and **20.3**

U.E.I.

2. A battery consists of six similar cells in series. Each cell has an e.m.f. of 2 V and an internal resistance of 0·1 Ω. The battery is connected to a lamp of resistance 3·4 Ω. Calculate the current supplied by the battery and the p.d. across the battery terminals.

Most relevant sections: 20.6 and **20.4.1**

U.E.I.

3. (a) Make a cross-sectional sketch of a lead–acid accumulator and clearly label the parts shown.
(b) What is meant if this accumulator is said to be of 50 A h capacity?

Most relevant sections: 20.3.1 and **17.1.1**

U.E.I.

4. A current of 20 A is supplied by a cell having an e.m.f. 2 V to a load with terminal voltage of 1·8 V. What is the internal resistance of the cell?

Most relevant section: 20.4.1

E.M.E.U.

5. What are the effects of local action and polarization in a simple cell?

Most relevant section: 20.2.1

U.E.I.

6. Sketch the construction of a 'dry' Leclanché cell and briefly explain the function of each part.

Most relevant section: 20.2.2

Y.C.F.E.

7. Sketch and clearly label a simple voltaic cell.

Most relevant section: 20.2.1

Y.C.F.E.

8. A certain type of dry cell has an e.m.f. of 1·5 V and internal resistance 0·6 Ω. What would be the e.m.f. and the internal resistance of a battery consisting of three such cells connected (a) in series, (b) in parallel?

Most relevant sections: 20.4.1, 20.6, and 20.7

C.G.L.I.

9. Explain what is meant by 'open-circuit e.m.f.', 'terminal potential difference' and 'internal resistance' as applied to an electric cell.
A cell which has an e.m.f. of 1·45 V and an internal resistance of 0·40 Ω is connected to an external circuit and supplies a current of 0·35 A to that circuit. Calculate:
(a) the terminal potential difference of the cell,
(b) the resistance of the external circuit.

Most relevant sections: 17.3, 20.4, and 20.4.1

E.M.E.U.

10. Four accumulators, each having an e.m.f. of 2 V and an internal resistance of 0·1 Ω, are used to supply a load of 3 Ω.
What would be the current flowing through the resistor and the potential difference across the resistor if the accumulators are: (a) in series, (b) in parallel?

Most relevant sections: 20.4.1, 20.6, and 20.7

U.E.I

11. Calculate the resistance of a group of three coils in parallel, having resistances of 40 Ω, 35 Ω, and 30 Ω respectively. If the group is connected across a 24-V battery having an internal resistance of 0·2 Ω, what will be the total current flowing and the current through the 35 Ω coil?

Most relevant sections: 17.8 and 20.4.1

U.C.L.I.

12. A circuit consists of two resistors of 6 Ω and 4 Ω respectively in parallel connected in series with a 10-Ω resistor. The resistors are supplied with current from a 12-V battery which has an internal resistance of 1 Ω.
Determine
(a) the current in the 4-Ω resistor,
(b) the p.d. across the 10-Ω resistor.
Sketch a diagram of the circuit and show on it the position in which an ammeter and a voltmeter would be placed to check the above calculated values.

Most relevant sections: 17.7, 17.8, 20.4.1, and 17.6.1

U.E.I.

13. A battery of e.m.f. 6 V and internal resistance 0·6 Ω is connected through a switch to two resistors of 14 Ω and 21 Ω in parallel. Calculate:
(a) the current passing through the switch,
(b) the p.d. across the resistors,
(c) the current passing through each resistor.
Make a diagram of the circuit, and state what instruments you would use, showing on your diagram where you would connect them, in order to make an experimental check of each of the above answers.

Most relevant sections: 17.8, 20.4.1, 17.5, and 17.6.1

C.G.L.I.

14. (a) What is the basic difference between primary and secondary cells?
(b) Explain the terms 'open-circuit e.m.f.', 'terminal p.d.' and 'internal resistance' as applied to an electric cell.
(c) A lead–acid cell is overcharged by 5 A for 10 h. If 1 C of electricity decomposes 0·1 mg of water find the amount of water required to be added to compensate for gassing.

Most relevant sections: 20.2, 20.3, and 17.1

U.E.I.

15. (a) What is the basic difference between *primary* and *secondary* cells?
(b) A lead–acid battery is charged at a constant charging rate of 3 A and is fully charged after 20 h. Calculate the quantity of electricity taken from the supply.
The ampere-hour capacity of the battery is 45 A h at the 5 A discharge rate. For how long would this battery supply a current of 5 A?
(c) Four resistors of 5 Ω, 8 Ω, 12 Ω, and 15 Ω are all connected in series with a battery. A voltmeter is connected across the 12-Ω resistor and indicates 6 V. Determine the terminal p.d. of the battery.

Most relevant sections: 20.2, 20.3, 17.1, 17.5, and 17.7

U.E.I.

ANSWERS TO EXERCISES 20
2. 3 A, 10·2 V. **4.** 0·01 Ω. **8.** (a) 4·5 V, 1·8 Ω, (b) 1·5 V, 0·2 Ω. **9.** (a) 1·31 V, (b) 3·75 Ω. **10.** (a) 2·35 A, 7·05 V, (b) 0·661 A, 1·983 V. **11.** 11·5 Ω, 2·05 A, 0·675 A. **12.** (a) 0·895 A, (b) 8·95 V. **13.** (a) 0·67 A, (b) 5·6 V, (c) 0·4 A, 0·266 A. **14.** (c) 18 g. **15.** (b) 60 A h, 9 h, (c) 20 V.

21 Electrical energy and power

21.1. ELECTRICAL ENERGY

Anything capable of doing work is said to possess energy, one form of which is electrical energy. By employing some suitable piece of apparatus electrical energy may be converted into other forms of energy, e.g. mechanical energy can be acquired by means of an electric motor, heat energy can be obtained by expending electrical energy in a resistor.

If one coulomb of electricity is caused to flow in a conductor by an e.m.f. of one volt then the amount of work done is one joule. The work done is equivalent to the energy consumed so that one joule of electrical energy is expended.

$$\text{Electrical energy } W = VQ = VIt. \tag{1}$$

Now, according to Ohm's law, $V = IR$ so that the equation $W = VIt$ becomes

$$W = I^2Rt. \tag{2}$$

Again, from Ohm's law $I = V/R$ and the equation $W = VIt$ can be written as

$$W = \frac{V^2t}{R}. \tag{3}$$

21.2. HEATING EFFECT OF ELECTRIC CURRENT

When a current flows in a conductor the material of which the conductor is made offers some measure of resistance to the flow, and the energy that must be used in overcoming the resistance is converted into heat. Experiments have shown that the amount of heat produced is directly proportional to:

(1) the *square of the current* (I^2),
(2) the *resistance* (R), and
(3) the *time* of current flow (t).

Unit resistance is that in which a current of one ampere flowing for one second produces one joule of heat energy. Hence when a current I ampere flows for t second through a resistance R ohm,

$$\text{heat energy produced} = I^2Rt. \tag{1}$$

It will be noted that this equation is the same as eqn (2) in section 21.1. It then follows that the amount of heat energy produced by an electrical current can also be obtained from:

$$\text{heat energy} = VIt \tag{2}$$

and
$$\text{heat energy} = \frac{V^2t}{R}. \tag{3}$$

These are eqns (1) and (3) from section 21.1.

21.3. ELECTRICAL POWER

Power is the rate at which work is done, i.e. the measure of the use or conversion of energy per unit time.

It has been seen that the heat energy produced when a current I ampere flows for t second through a resistance R ohm is I^2Rt joule.
Then,

$$\text{electrical power } P = \frac{I^2Rt}{t}$$
$$= I^2R. \tag{1}$$

From section 21.2 it follows that electrical power is also given by:

$$P = \frac{V^2}{R}. \tag{2}$$

and

$$P = VI. \tag{3}$$

When dealing with large electrical plant and machinery the watt is of inconvenient magnitude so that the multiple units of kilowatt (1 kW = 1000 W) and megawatt (1 MW = 1 000 000 W) are used. In the same way the joule, or watt-second, is inconveniently small as the unit of electrical energy and so the watt-hour (1 W h = 3600 J) and kilowatt-hour (1 kW h = 3 600 000 J) are employed. The kilowatt-hour is also known as a Board of Trade unit, or simply, a 'unit'.

21.4. COST OF ELECTRICAL ENERGY

Since electrical energy cannot be stored (except on d.c. systems where a certain amount of storage is possible by using batteries) the methods by which consumers are charged for the energy used may appear rather complex. In addition to a charge proportional to the amount of electrical energy used, some share of the cost of the generating plant and distribution system must be borne by consumers.

Various tariff systems are in use, although it will be sufficient here to consider only the price per 'unit'. Thus if the charge is say 5p/kW h, and a 3-kW immersion water heater is in use for 4 h, the energy consumed is 12 kW h and the cost will be 12 kW h × 5p = 60 p.

21.5. EFFICIENCY OF CONVERSION OF ENERGY

It has previously been stated that in all operations involving the conversion of energy into work there is some waste. The process of converting heat energy into mechanical energy was seen to be unavoidably wasteful, a large proportion of the heat being 'lost'. With electrical machinery and apparatus the efficiency of conversion is much higher. The following are given as typical examples: electric motors (electrical to mechanical energy), generators (mechanical to electrical energy), and electric heaters (electrical to heat energy) have efficiencies of the order of 80 to 90 per cent.

The measure of the usefulness of an operation, or machine, is termed its efficiency.

$$\text{Efficiency} = \frac{\text{work (or power) output}}{\text{work (or power) input}}.$$

For convenience, efficiency is generally expressed as a percentage.

EXAMPLE 21.1. A d.c. motor with a series resistor of 10 Ω is used on a 240-V mains supply. When the motor is running, the p.d. across the series resistor is found to be 40 V.

(a) Find the p.d. across the motor and the current passing through it.
(b) Find the power, in kW, taken from the mains supply.
(c) Find the power, in kW, taken by the resistor and converted to heat.
(d) If 55 per cent of the work input to the motor is converted to mechanical output, what power is developed by the motor?

C.G.L.I.

SOLUTION

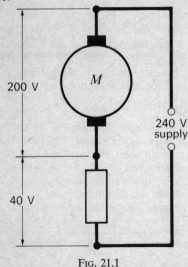

FIG. 21.1

(a) In a series circuit, the sum of the separate p.d.s is equal to the e.m.f. (see Fig. 21.1). Thus the p.d. across the motor

$$= \text{e.m.f.} - \text{p.d. across resistor}$$
$$= (240 - 40) \text{ V}$$
$$= 200 \text{ V}.$$

Current through motor = current through resistor (since they are in series).

$$\therefore I = \frac{V(\text{i.e. p.d. across resistor})}{R \ (\text{i.e. resistance of resistor})}$$
$$= \frac{40}{10} = 4.$$

Current = 4 A.

(b) Power taken from mains supply,

VI (where V is the supply voltage)
$$= 240 \times 4 = 960.$$
$$\text{Power} = 960 \text{ W} = 0.960 \text{ kW}.$$

(c) Power taken by resistor.

VI (where V is the p.d. across the resistor)
$$= 40 \times 4 = 160$$
$$\text{Power} = 160 \text{ W} = 0.160 \text{ kW}.$$

Or, power = $(I^2 R)$
$$= (4)^2 \times 10 \text{ W} = 160 \text{ W} = 0.160 \text{ kW}.$$

Or, power $\left(= \frac{V^2}{R}\right) = \frac{(40)^2}{10} = \frac{1600}{10} \text{ W}$
$$= 160 \text{ W} = 0.160 \text{ kW}.$$

(d) Power input to motor = power from mains − power taken by resistor
$$= (960 - 160) \text{ W}$$
$$= 800 \text{ W}.$$

Power output of motor = power input to motor × efficiency
$$= 800 \times \frac{55}{100} \text{ W}$$
$$= 440 \text{ W}$$
$$= 0.44 \text{ kW}.$$

EXAMPLE 21.2. An electrical immersion heater has a loading of 2½ kW and is used to heat a tank containing 180 l of water. If the temperature of the water was originally at 10 °C (283 K) and it is to be raised to 75 °C (348 K), find the time taken assuming the efficiency of the tank and heater is 75 per cent. How much would it cost to heat the water at 6 p/kW h? Take the specific heat capacity of water as 4·2 kJ/kg K.

Y.C.F.E.

SOLUTION

Mass of 1 l of water = 1 kg.

$$\begin{array}{l} \text{Quantity of heat} \\ \text{absorbed by water} \end{array} = \begin{array}{l} \text{mass of} \\ \text{water} \end{array} \times \begin{array}{l} \text{specific} \\ \text{heat} \\ \text{capacity} \end{array} \times \begin{array}{l} \text{tempera-} \\ \text{ture rise} \end{array}$$

$$= 180 \times 4 \cdot 2 \times (75 - 10) \text{ kJ}$$
$$= 49\,140 \text{ kJ}.$$

But the overall efficiency is 75 per cent, so that heat energy taken from supply

$$= 49\,140 \times \frac{100}{75} \text{ kJ}$$
$$= 65\,520 \text{ kJ}.$$

Taking 3600 kJ as equivalent to 1 kW h,

$$\text{electrical energy input} = \frac{65\,520}{3600} \text{ kW h}.$$

Time taken by $2\frac{1}{2}$ kW heater

$$= \frac{\text{electrical energy (kW h)}}{\text{electrical power (kW)}}$$
$$= \frac{65\,520}{3600 \times 2\frac{1}{2}}$$
$$= 7 \cdot 25 \text{ h}.$$

$$\text{Cost} = \text{energy} \times 6 \text{ p/kW h}$$
$$= \frac{65\,520}{3600} \times 6$$
$$= 109 \cdot 2 \text{ p, say } £1 \cdot 10.$$

EXAMPLE 21.3. A steam turbine drives a generator which has an output of 20 000 kW. The fuel burnt to produce the steam has a calorific value of 33·5 MJ/kg. What is the rate of fuel consumption if the overall efficiency of the plant is 20 per cent?

N.C.T.E.C.

SOLUTION

$$\text{Generator output} = 20\,000 \text{ kW}$$
$$= 20 \text{ MW}.$$

Overall efficiency of plant is 20 per cent, so that

$$\text{input is equivalent to } \frac{\text{output}}{\text{efficiency}}$$
$$= \frac{20 \text{ MW}}{20/100}$$
$$= 100 \text{ MW}.$$

Now　　　1 MW = 1 MJ/s,

∴ heat input = 100 MJ/s,

and rate of fuel consumption

$$= \frac{100 \text{ MJ/s}}{33 \cdot 5 \text{ MJ/kg}}$$
$$= 3 \text{ kg/s}.$$

SUMMARY

Electrical energy $W = VQ = VIt$
$$= I^2 Rt$$
$$= \frac{V^2 t}{R}.$$

The unit of electrical energy is the joule (or watt-second)

$$1 \text{ W h} = 3600 \text{ J},$$
$$1 \text{ kW h} = 3\,600\,000 \text{ J}.$$

The heat produced by an electric current flowing through a resistor is proportional to:

(1) the square of the current (I^2),
(2) the resistance (R),
(3) the time of current flow (t).

$$\text{Heat energy} = I^2 Rt$$
$$= VIt$$
$$= \frac{V^2 t}{R} \text{ (joules)}.$$

Electrical power $P = I^2 R$
$$= \frac{V^2}{R}$$
$$= VI.$$

The unit of electrical power is the watt (or joule per second).

$$1 \text{ kW} = 1000 \text{ W},$$
$$1 \text{ MW} = 1\,000\,000 \text{ W}.$$

$$\text{Efficiency} = \frac{\text{work (or power) output}}{\text{work (or power) input}}.$$

Efficiency has no units, but is expressed as either a percentage or compared to unity.

EXERCISES 21

1. A shaping machine is driven by an electric motor which delivers an output of 2·25 kW. The motor, which is 90 per cent efficient, operates on a 240-V supply. Determine:
   (a) the input power to the motor,
   (b) the current consumption;
   (c) the average work done per minute.

Most relevant sections: **21.3** and **21.5**

U.E.I.

2. Define efficiency in terms of either work or power or energy. Determine the efficiency of conversion of a 2-kW heater which raises 100 kg of water through 60 °C in 4 h. Specific heat capacity of water = 4·2 kJ/kg K.

Most relevant sections: **21.5**, **21.2**, and **12.1**

Y.C.F.E.

3. On full load and developing 22 kW, the efficiency of an electric motor is 88 per cent. If the input voltage is 240 d.c. calculate the input power and current.

Most relevant sections: **21.5** and **21.3**

Y.C.F.E.

4. State the electrical unit of power.
   Calculate the cost per hour of running a plant which consumes 300 kW if the cost of electrical energy is 6 p/kW h.

Most relevant section: **21.3**

E.M.E.U.

5. A 20-Ω resistance is connected to a 200-V d.c. supply. Calculate the current flowing and the quantity of heat which is generated per minute.

Most relevant sections: **17.5** and **21.2**

E.M.E.U.

6. Calculate the cross-sectional area of the wire of a generator field coil which is wound from 300 m of copper wire having resistance 20 Ω. The resistivity of copper is 17 $\mu\Omega$ mm. When the coil is connected to a 12 V d.c. supply for 5 min, determine:

   (a) the current in the coil,
   (b) the power loss, and
   (c) the energy loss.

Most relevant sections: **17.9.3**, **17.5**, **21.3**, and **21.1**

Y.C.F.E.

7. An electric motor drives a crane lifting a mass of 5 t at a speed of 0·3 m/s. The efficiency of the crane is 60 per cent and the efficiency of the motor 85 per cent. Calculate:

   (a) the output of the motor (kW);
   (b) the input of the motor (kW);
   (c) the electrical energy used by the motor in lifting the load through 15 m; and
   (d) the current taken from the 240-V supply.
   ($g = 9\cdot81$ m/s^2.)

Most relevant sections: **7.3**, **21.5**, and **21.3**

U.L.C.I.

8. An electric train of mass 250 t travels at a uniform speed of 65 km/h on a horizontal track against a resistance of 55 N/t. Calculate the power exerted by the motors. If the supply voltage is 1500 V calculate the current taken from the supply
   Gearing efficiency = 80 per cent.
   Motor efficiency = 85 per cent.

Most relevant sections: **7.3**, **21.5**, and **21.3**

U.L.C.I.

9. (a) It is required to deposit, from a copper sulphate solution, a coating of copper 0·06 mm on to a plate of total surface area 120 cm^2. If the supply voltage is 6 V and the circuit resistance is 3 Ω, for how long must the current flow through the electrolyte? Take the electrochemical equivalent and density of copper as 0·000 33 g/C and 8·8 g/cm^3 respectively.
   (b) Calculate the electrical energy used in coating 2000 identical plates.

Most relevant sections: **19.4** and **21.1**

U.E.I.

10. In a hydro-electric power station water falls at the rate of 2·3 m^3/s from a height of 120 m. If the turbine and generator efficiencies are 65 per cent and 78 per cent respectively, calculate:
    (a) the output of the turbine in megawatts,
    (b) the output of the generator in megawatts, and
    (c) the current supplied by the generator, if the output voltage is 6600 V.
    Density of water = 1000 kg/m^3; $g = 9\cdot81$ m/s^2.

Most relevant sections: **7.3**, **21.5**, and **21.3**

C.G.L.I.

11. An electric furnace rated at 3 kW has an efficiency of 90 per cent and is used to heat 15 kg of steel. Calculate the increase of temperature per minute, assuming the specific heat capacity of steel to be 0·46 kJ/kg K.

Most relevant sections: **21.2**, **21.5**, and **12.1**

E.M.E.U.

12. An electric furnace has a rating of 3 kW and a thermal efficiency of 80 per cent. The furnace is loaded with $2\frac{1}{2}$ kg of steel rods of specific heat capacity 0·462 kJ/kg K at a temperature of 20 °C. To what temperature are they heated if the time of operation is 7 min?

Most relevant sections: **21.2**, **21.5**, and **12.1**

U.E.I.

13. An oil engine with an efficiency of 23 per cent drives a generator. The generator which develops 5 kW at 250 V is itself only 75 per cent efficient. If the fuel used by the engine has a calorific value of 44·2 MJ/kg, determine:

   (a) the power output of the engine,
   (b) the engine fuel consumption in kg/h,
   (c) the generator load current, and
   (d) the overall efficiency of the unit.

Most relevant sections: 21.3, 21.2, and 21.5

U.E.I.

14. A 200-V d.c. motor has a power output of 1·5 kW and is driving a pump whose efficiency is 60 per cent. The power input to the motor is 1·8 kW. Determine
   (a) the current taken by the motor,
   (b) the efficiency of the motor,
   (c) the pump output,
   (d) the overall efficiency of the plant.

Most relevant sections: 21.3 and 21.5

U.E.I.

15. A circuit consisting of two resistors A and B joined in parallel, is connected in series with a third resistor C across a 12-V d.c. supply. The resistors have the following resistances: $A = 3\,\Omega$, $B = 6\,\Omega$, and $C = 2\,\Omega$. Sketch the circuit diagram and calculate the current flowing through, the p.d. across, and the power dissipated in resistor C.

Most relevant sections: 17.5, 17.7, 17.8, and 21.3

U.E.I.

16. (a) Why is the mechanical energy used to drive a generator greater than the electrical energy obtained from its terminals?
   (b) An oil engine drives a d.c. generator which has a full-load output of 150 kW at 500 V. Calculate the current supplied by the generator when fully loaded.

   The fuel oil used has a calorific value of 46 MJ/kg. What is the rate of fuel consumption if the overall efficiency of the plant is 20 per cent?

Most relevant sections: 21.3, 21.5, and 12.4

U.E.I.

ANSWERS TO EXERCISES 21

1. (a) 2·5 kW, (b) 10·4 A, (c) 135 MJ. 2. 87·5 per cent. 3. 25 kW, 104 A. 4. £18. 5. 10 A, 120 kJ. 6. 0·255 mm², (a) 0·6 A, (b) 7·2 W, (c) 2160 J. 7. (a) 24·5 kW, (b) 28·8 kW, (c) 0·4 kW h, (d) 120 A. 8. 309·4 kW, 242 A. 9. (a) 2 h 40 min, (b) 64 kW h. 10. (a) 1·76 MW, (b) 1·37 MW, (c) 208 A. 11. 23·5 °C (296·5 K). 12. 895 °C (1168 K). 13. (a) 6·7 kW, (b) 2·36 kg/h, (c) 20 A, (d) 17·25 per cent. 14. (a) 9 A, (b) 83·3 per cent, (c) 0·9 kW, (d) 50 per cent. 15. 3 A, 6 V, 18 W. 16. (b) 300 A, 16·3 g/s.

22 Magnetism and electromagnetism

22.1. MAGNETIC MATERIALS

Magnetism is an effect produced by moving electrons in certain materials. Iron, nickel, and cobalt are the only known elements which possess definite magnetic properties; of these iron is outstanding. Various materials with similar magnetic properties to iron are known as ferromagnetic materials and these experiences force acting on them when placed near a magnet.

Magnets can be classified as:

(1) *permanent magnets*, which retain their magnetic properties permanently, or
(2) *electromagnets*, which possess magnetic properties due to the presence of an electric current.

Tungsten and cobalt steels were once the chief materials from which permanent magnets were made, but in recent years the use of iron alloys containing aluminium, nickel, and cobalt has enabled considerable reductions to be brought about in the size of permanent magnets for many purposes. 'Alnico' (Al, Ni, Co, Fe) and 'Alni' (Al, Ni, Fe) are alloys commonly used in the manufacture of very strong permanent magnets. Mechanically these alloys are intensely hard, impossible to forge, and difficult to machine except by grinding.

22.2 MAGNETIC FIELDS

A *magnetic field of flux* is the region in which the forces of a magnet can be discovered. The field surrounds a magnet in all directions, the field strength being greatest near two points on the magnet known as the *poles*. A magnetic field is mapped by an arrangement of *lines* giving an indication of the strength and direction of the flux, although it should be remembered that these lines are used for convenience only and in reality do not exist.

When freely suspended horizontally a magnet sets itself N–S parallel to the earth's magnetic field. This is due to the fact that unlike magnetic poles attract one another; similarly, like poles will repel. In a given magnet the two poles are of equal strength and are situated near the ends of the magnet.

Figs 22.1–22.6 show the conventions of polarity and direction of flux adopted.

22.3. MAGNETIC EFFECT OF AN ELECTRIC CURRENT

Whenever an electric current flows in a conductor a magnetic field, in the form of concentric circles, is set up

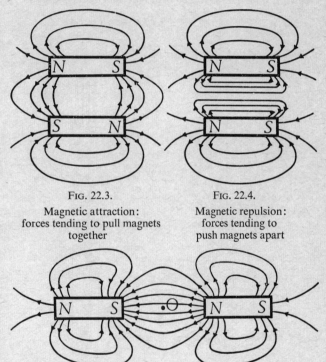

FIG. 22.3.
Magnetic attraction: forces tending to pull magnets together

FIG. 22.4.
Magnetic repulsion: forces tending to push magnets apart

FIG. 22.5. Magnetic attraction: combined field strongest at point 0

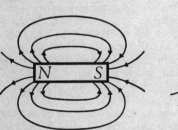

FIG. 22.1.
Field of bar magnet

FIG. 22.2.
Field of horse-shoe magnet

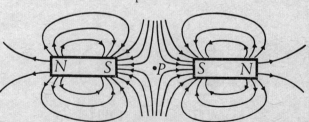

FIG. 22.6. Magnetic repulsion: fields exactly equal and opposite at 'neutral point' P

around the conductor. The field is present along the whole length of the conductor and the intensity of the flux, which is proportional to the magnitude of the current, is greatest near the conductor. If the direction of the current flow is reversed, a corresponding reversal of the magnetic field of flux is obtained.

It has previously been stated that a magnetic field is mapped by an arrangement of lines. Fig. 22.7 shows the pattern of the field around a straight current-carrying conductor. The direction of the field can be determined from a knowledge of the direction of the current flow and by applying the *right-hand screw rule*.

Fig. 22.8 shows the comparison between a right-hand screw and the direction of a magnetic field due to a flow of current. The conventional method of indicating the current direction at the ends of a conductor is shown in Fig. 22.9.

If a current-carrying conductor is bent into a single-turn loop, the magnetic field set up by the current will follow the form of the loop as is indicated in Fig. 22.10. By applying the right-hand screw rule the direction of the flux is again determined from a knowledge of the direction of current flow.

Now when a long conductor is wound in the same direction to form a multi-turn coil, and a current is passed through the coil, all the turns assist one another in producing a strong magnetic field. Such an arrangement is known as a *solenoid*. In practice, the turns are tightly wound side by side, layer upon layer, and the magnetic field of flux is intensified by the insertion of a ferro-magnetic core inside the solenoid. The resulting device is an electromagnet, the core of which behaves like a permanent magnet, with its own magnetic poles, for as long as the current is maintained in the solenoid.

The principle of a solenoid is illustrated in Fig. 22.11, where for convenience only a few turns are shown. A useful method of ascertaining the polarity of a solenoid is given, where the part end-views of the solenoid are shown in third angle projection.

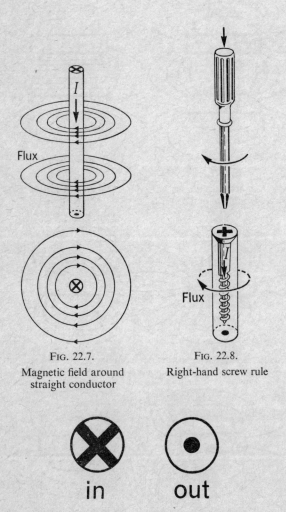

FIG. 22.7.
Magnetic field around straight conductor

FIG. 22.8.
Right-hand screw rule

FIG. 22.9. Convention for direction of current

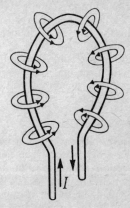

FIG. 22.10. Magnetic field around single-turn loop

22.4. CONDUCTOR IN A MAGNETIC FIELD

The magnetic field between dissimilar poles is shown in Fig. 22.5, and the field due to an electric current flowing in a conductor is illustrated in Fig. 22.7. Now consider a straight conductor lying at right-angles to a magnetic field such that its length lies within the field. When a current is passed through the conductor the two fields will interact in such a way that a force will be set up on the conductor which will tend to move it. This will be readily understood from a study of Fig. 22.12, which shows the separate fields. It will be seen that in the spaces above and below the conductor, the fields oppose and assist one another respectively. As a result the interacting field is weakened above and strengthened below the conductor causing the force F to act on the conductor in the direction shown in Fig. 22.13. This phenomenon is

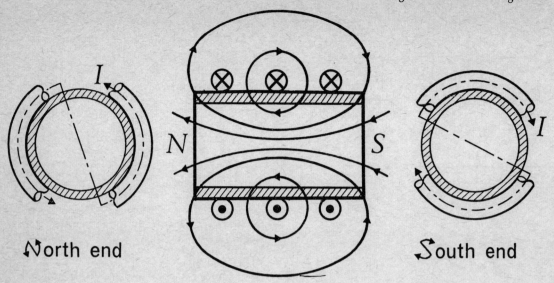

North end　　South end

FIG. 22.11. Magnetic field of a solenoid

known as the *motor* effect of an electric current. If the direction of the current flow is reversed, a corresponding reversal in the direction of the force F on the conductor is obtained.

The direction in which the force F acts depends upon:

(i)　the direction of the magnetic field of flux, and
(ii)　the direction of the current flow.

All three directions are at right-angles to each other and any one direction can be determined from a knowledge of the other two by the use of *Fleming's left-hand (motor) rule* as illustrated in Fig. 22.14.

The left-hand (motor) rule states that:

When the thumb and first two fingers of the left hand are held at right angles to each other in such a way that the first finger is pointed in the direction of the magnetic field of flux, and the second finger is pointed in the direction of the current flow, then the thumb will be pointed in the direction of the force F acting on the conductor.

The magnitude of the force F on a current-carrying conductor lying at right-angles to a magnetic field of flux is proportional to three quantities:

(i)　the *density* (B) of the magnetic flux,
(ii)　the *magnitude* of the *current* (I) flowing, and
(iii)　the *effective length* (l) of the conductor lying in the field.

Thus the force $F \propto BIl$,

and $F = kBIl$, where k is a constant.

Now, by using SI units, and choosing a suitable unit for the flux density, the constant k can be made equal to 1 and can be omitted.

If the current (I) through the conductor $= 1$ A, the effective length (l) of the conductor $= 1$ m, and the force (F) on the conductor $= 1$ N,
then, the flux density (B) $= 1$ tesla (T).

It therefore follows that the force on any current-carrying conductor lying in and at right-angles to a magnetic field of flux is given by:

$$F = BIl \text{ (newtons).}$$

The total flux (Φ) in a magnetic field is measured in webers (Wb), and

$$\text{flux density} = \frac{\text{total flux}}{\text{area of field}} \text{ (teslas),}$$

or
$$B(\text{T}) = \frac{\Phi \text{ (Wb)}}{A \text{ (m}^2)}.$$

In practice, values of flux density are generally small, and the submultiple units of millitesla (mT) and micro-tesla (μT) are often used.

$$1 \text{ mT} = 10^{-3} \text{ T,}$$
$$1 \text{ } \mu\text{T} = 10^{-6} \text{ T.}$$

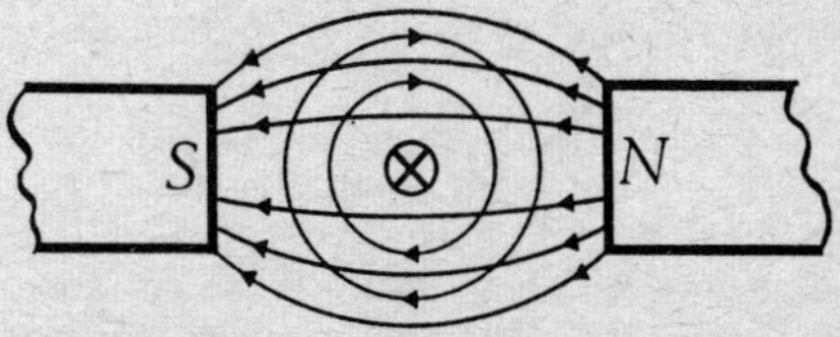

FIG. 22.12. Interacting fields shown separately

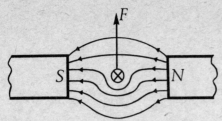

FIG. 22.13. Resultant interacting field

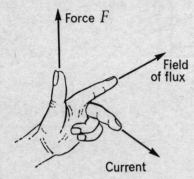

FIG. 22.14. Fleming's left-hand rule for motors

22.4.1. Force between parallel conductors. It was shown in Fig. 22.7 that a current-carrying conductor is surrounded by a magnetic field, and in Fig. 22.13 that a current-carrying conductor placed in and at right-angles to a magnetic field has a force acting upon the conductor. It therefore follows that when two current-carrying conductors are placed parallel to each other, there is a force acting on each of the conductors as shown in Fig. 22.15.

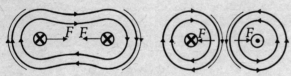

FIG. 22.15. Interacting fields due to two parallel current-carrying conductors

The force is used in the *international definition of the ampere*. This definition states that:

The ampere is that constant current which, when flowing in each of two straight infinitely long parallel conductors (of negligible cross-section) placed in a vacuum and separated 1 metre between centres, causes each conductor to have acting upon it a force of 2×10^{-7} newtons per metre length.

EXAMPLE 22.1. A straight conductor lies at right-angles to a magnetic field of flux density 1·5 T such that 200 mm of its length lies within the field. If the conductor forms part of a closed circuit having a total resistance of 0·1 Ω and p.d. 1·8 V, calculate the force on the conductor.

SOLUTION

(*Note:* All quantities must be measured in basic units). Force on conductor

$$F = BIl,$$

where flux density $B = 1.5$ T,

$$\text{current } I = \frac{V}{R} = \frac{1.8}{0.1} = 18 \text{ A},$$

and length $l = 200$ mm $= 0.2$ m.

$$\therefore \text{ force } F = 1.5 \times 18 \times 0.2$$
$$= 5.4.$$

Force conductor $= 5.4$ N.

EXAMPLE 22.2. A rectangular coil wound with 50 turns of wire has a mean width of 15 mm and a mean axial length of 20 mm. When situated in a uniform magnetic field of flux density 200 mT, the current in the coil is 10 mA. Calculate:
(a) the force acting on one side of the coil, and
(b) the maximum torque on the coil.

SOLUTION

(*Note:* The coil is wound with 50 turns so that each side of the coil is equivalent to 50 conductors each 20 mm long. The ends of the coil are not across the field and no force is exerted on them. All quantities must be measured in basic units.)

(a) Force per conductor $F = BIl$,

where flux density $B = 200$ mT $= 0.2$ T,

current $I = 10$ mA $= 0.01$ A,

and length $l = 20$ mm $= 0.02$ m.

$$\therefore \text{ force } F = 0.2 \times 0.01 \times 0.02 \text{ N}.$$

Force on conductor $= 0.2 \times 0.01 \times 0.02$ N.

Then, force on one side of coil (i.e. on 50 conductors)

$$= 50 \,(0.2 \times 0.01 \times 0.02) \text{ N}$$
$$= 0.002 \text{ N}$$
$$= 2 \times 10^{-3} \text{ N}.$$

(b) Torque $(T) = \text{force} \times \text{radius}$,

where radius $= 7.5$ mm $= 0.007\,5$ m.

$$\therefore T = 0.002 \times 0.007\,5$$

Torque $= 0.002 \times 0.007\,5$ N m.

But the coil has two sides,

$\therefore$ maximum torque on coil

$$= 2 \,(0.002 \times 0.007\,5) \text{ N m}.$$
$$= 0.000\,03 \text{ N m}$$
$$= 3 \times 10^{-5} \text{ N m}.$$

22.5. ELECTROMAGNETIC INDUCTION

Whenever relative motion occurs between a magnetic field of flux and a conductor at right-angles to that flux, an e.m.f. is induced, or generated, in the conductor. If the conductor forms part of a closed circuit, a current will flow in a direction corresponding with the direction of the induced e.m.f., and a force will be set up on the conductor opposing the motion. The direction of the e.m.f. depends upon:

(1) the direction of the magnetic field of flux, and
(2) the direction in which the conductor moves relative to the field,
and can be determined by two methods as follows:
(a) *Lenz's law* (see Fig. 22.16) which states that:

The direction of an induced e.m.f. is such that it tends to set up a flow of current, which in turn causes a force opposing the motion which is generating the e.m.f.

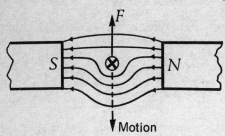

FIG. 22.16. Lenz's law

(b) *Fleming's 'right-hand' (generator) rule* (see Fig. 22.17), which states that:

When the thumb and first two fingers of the right hand are held at right-angles to each other in such a way that the first finger is pointed in the direction of the magnetic field of flux, and the thumb is pointed in the direction of motion, or relative motion, of the conductor, then the second finger will be pointed in the direction of the induced e.m.f.

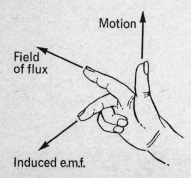

FIG. 22.17. Fleming's right-hand rule for generators

The unit of magnetic flux can be defined as follows:
The weber is that magnetic flux which, when cut at an unvarying rate by a conductor in 1 second, induces an e.m.f. of 1 volt.

Hence, the magnitude of the induced e.m.f. is dependent upon the rate at which the conductor cuts the flux, and is given by:

$$\textbf{induced e.m.f.} = \frac{\textbf{flux}}{\textbf{time}} \textbf{ (Volts)},$$

or

$$E(\text{V}) = \frac{\Phi}{t} \frac{(\text{Wb})}{(\text{s})}.$$

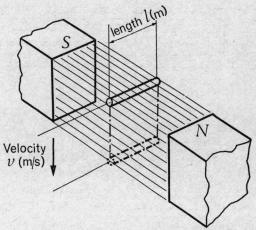

FIG. 22.18. Conductor cutting magnetic field of flux

From Fig. 22.18,
area of flux cut by conductor in 1 s, $A = lv\,(\text{m}^2)$, and total flux cut per second, $\Phi = BA = Blv$ (Wb).
But, flux cut per second = induced e.m.f.
$$\therefore \text{ induced e.m.f. } E = Blv \text{ (V)}.$$

Where a conductor moves such that the direction of motion makes an angle θ with the plane of the flux, the velocity v is resolved into two components (Fig. 22.20) $v \cos \theta$ parallel to the flux, and $v \sin \theta$ perpendicular to the flux. The magnetic flux is crossed by the component $v \sin \theta$ which gives rise to an e.m.f. of $Blv \sin \theta$.

$$\therefore \textbf{ induced e.m.f. } E = Blv \sin \theta.$$

It will be seen that a maximum e.m.f. is generated when the conductor is moving at right-angles to the plane of the flux (Example 22.3, Fig. 22.19), and that no e.m.f. is induced when the conductor is moving parallel with the flux.

EXAMPLE 22.3. A coil of 200 turns, with a coil side of active length 300 mm, rotates in a uniform magnetic field between the poles of an electromagnet which give a field of flux density 0·02 T. If the conductor speed is 12·6 m/s. determine:

(a) the maximum e.m.f. generated in the coil, and
(b) the e.m.f. generated at the instant when the conductors are 30° from maximum position

Y.C.F.E.

SOLUTION

(a) (*Note:* Maximum e.m.f. is generated when the plane of the coil is in the plane of the flux, as shown in Fig. 22.19.)

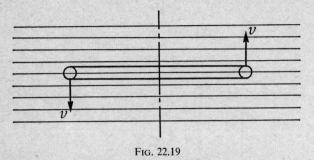

Fig. 22.19

Induced e.m.f. per conductor, $E = Blv$,

where flux density $B = 0.02$ T,

length $l = 300$ mm $= 0.3$ m,

and velocity $v = 12.6$ m/s.

$$\therefore E = 0.02 \times 0.3 \times 12.6 \text{ V}.$$

Induced e.m.f. per conductor $= 0.02 \times 0.3 \times 12.6$ V.

But the coil has 200 turns which is equivalent to 200 conductors per side, i.e. 400 conductors.

Then, maximum e.m.f. generated in coil

$$= 400 \, (0.02 \times 0.3 \times 12.6) \text{ V}$$
$$= 30.24 \text{ V}.$$

(b) (*Note:* When the conductors are 30° from maximum position, $\theta = 60°$ as shown in Fig. 22.20.)

Then e.m.f. generated in coil $= 400 \, (Blv \sin \theta)$ V.

But 400 $Blv = 30.24$ (part (a)),

$$\therefore 400(Blv \sin \theta) = 30.24 \sin 60°$$
$$= 30.24 \times 0.866$$
$$= 26.19.$$

Thus e.m.f. generated in coil $= 26.19$ V.

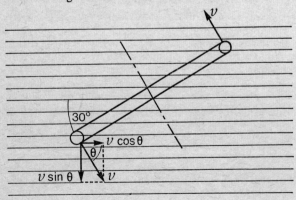

Fig. 22.20

22.6. MAGNETIC DEVICES

There are many different devices which operate by means of magnetism. These include motors, alternators (see Chapter 23), indicating instruments such as ammeters and voltmeters (see Chapter 24), electro-lifting magnets, transformers, chokes for discharge tube lighting, loud-speakers, cathode-ray-tube focusing in television receivers, clocks, relays, telephone receivers, and magnetic chucks for workholding (see section 22.6.1 below). The magnets employed are either permanent magnets or electro-magnets, the choice depending on the various design requirements, size, operating conditions, etc.

22.6.1. Magnetic workholding. Permanent magnet chucks are essentially mechanical workholding devices, relying on permanent magnets to provide the holding down of ferrous workpieces. The two principal components in magnetic workholding are (1) the workpiece and (2) the magnetic chuck, and the characteristics of both of these components will greatly influence the force of attraction experienced by the workpiece.

The holding force, F(N), exerted on a workpiece is given by the equation:

$$F = B^2 A/k,$$

where B is the flux density in the workpiece (T), A is the effective area of the flux path (m²), and k is a constant. It will be seen that, since $F \propto B^2 A$, the flux density at the holding face should be as high as possible, and the height of the workpiece and the effective area of the holding face should be as large as possible. The effective area is that area of the workpiece actually touching the surface of the chuck. Therefore the effective area not only depends on the area of the face of the workpiece, but also on the surface finish of this face (e.g. rough machined, finish ground, etc.)

A very strong magnetic alloy used in permanent magnet chucks is Alcomax III (Al, Co, Ni, Cu, Fe). The construction of a typical chuck is shown in Figs 22.21 and 22.22, and the principle is that the magnetic flux must be conducted through the workpiece in the 'on' position and diverted in the 'off' position. The magnets are assembled in a movable grid which is operated by means of an eccentric pin attached to the 'on–off' handle. The top plate includes inserts of non-magnetic white metal, and it is these inserts which direct the flux through

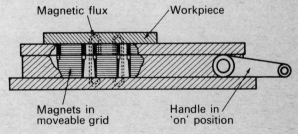

Fig. 22.21

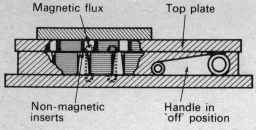

Magnetic flux   Top plate

Non-magnetic   Handle in
inserts   'off' position

FIG. 22.22

the workpiece in the 'on' position and allow the flux to be diverted in the 'off' position.

SUMMARY

Magnets can be classified as:

(1) permanent magnets or
(2) electromagnets.

Like magnetic poles repel one another: unlike poles attract one another.

$$\text{Flux density } B = \frac{\text{total flux } \Phi}{\text{area } A} \text{ (tesla)}.$$

Submultiple units of flux are:

$$1 \text{ mWb} = 10^{-3} \text{ Wb (milliweber)},$$

$$1 \ \mu\text{Wb} = 10^{-6} \text{ Wb (microweber)}.$$

A current-carrying conductor is surrounded all along its length by a magnetic field.

The force F on a current-carrying conductor lying in and at right-angles to a magnetic field of flux is given by

$$F = BIl \text{ (N)}.$$

The e.m.f. E induced in a conductor cutting a magnetic field at right-angles to the plane of flux is given by

$$E = \frac{\Phi}{t} = Blv \text{ (V)}.$$

Where a conductor moves such that the direction of motion makes an angle θ with the plane of the flux:

$$E = Blv \sin \theta \text{ (V)}.$$

(*Note:* Basic units must always be used.)

EXERCISES 22

1. Explain with the aid of a diagram what is meant by a 'neutral' point in a magnetic field.

Most relevant section: 22.2

E.M.E.U.

2. Fig. 22.23 shows a current-carrying conductor lying between a pair of magnetic poles. Sketch the resultant magnetic field and indicate its effect on the conductor.

Most relevant section: 22.4

U.E.I.

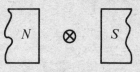

FIG. 22.23

3. Sketch the shape of the magnetic field produced by an electric current in (a) a straight wire (b) a single turn coil of wire. Indicate in each case the direction of current and of magnetic field.

Most relevant section: 22.3

C.G.L.I.

4. Sketch the magnetic field due to a current flowing in a solenoid. Indicate the direction of the current and mark the north pole of the solenoid.

Most relevant section: 22.3

Y.C.F.E.

5. Explain, with the aid of sketches, what happens when a current is passed through a conductor lying in a uniform magnetic field at right-angles to the lines of force.

Most relevant section: 22.4

E.M.E.U.

6. The force exerted on a conductor situated in a magnetic field depends upon three factors. State these and write down a relationship between them and the force exerted, naming the units used.

Most relevant section: 22.4

U.E.I.

7. A conductor 250 mm long and carrying a current of 75 A is placed at right angles to a magnetic field of flux density 0·80 T. Calculate the force on this conductor.

Most relevant section: 22.4

Y.C.F.E.

8. A straight conductor 40 mm long lies in and at right-angles to a magnetic field. When the conductor carries a current of 5 A the force acting on it is 0·1 N. Find the flux density of the magnetic field stating the unit of measurement.

Most relevant section: **22.4**

C.G.L.I.

9. Describe how you would construct an electromagnet given a suitable soft iron bar, insulated wire, and a source of p.d. Sketch the circuit and indicate, by means of the diagram, the polarity of the magnet in relation to the direction of current flow. Explain how you would test for polarity.

Most relevant section: **22.3**

E.M.E.U.

10. In Fig. 22.24 *abcd* is a rectangular coil wound with 50 turns of wire in which a current of 0·5 A flows in the direction shown.

The coil is in a magnetic field of uniform flux density = 0·25 T as indicated.

(a) Find the force acting on each of the sides of the coil *ab* and *cd* giving the magnitude and direction in both cases.

(b) If the coil were free to rotate about axis *XX* find the torque tending to turn it and say how far it would turn before the torque became zero, giving reasons.

Most relevant sections: **22.4** and **9.7**

C.G.L.I.

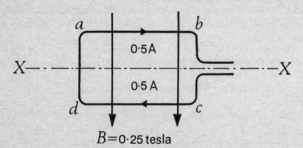

B=0·25 tesla

ab=5cm ; bc=3cm

Fɪɢ. 22.24

11. The unit of magnetic flux, the weber, may be defined as that magnetic flux which, when cut by a conductor in one second, generates an e.m.f. of 1 V.

Using the above definition, calculate the e.m.f. generated in a wire 250 mm long, which is moving at a

speed of 5 m/s at right-angles to its length in a magnetic field of 0·4 T. Show, with simple sketches, how the direction of the e.m.f. is determined by the direction of the magnetic field and the direction of motion of the conductor.

Most relevant section: **22.5**

U.E.I.

12. A generator conductor having a length of 200 mm at radius 100 mm is moving at 9·55 rev/s at right-angles to the magnetic field which has a flux density of 1·4 T. If the conductor current is 30 A, what is the e.m.f. induced, and what is the force on the conductor?

Most relevant sections: **22.5** and **22.4**

U.E.I.

13. A straight conductor 300 mm long is moved at 6 m/s in a direction at right-angles to its length and at right-angles to a uniform magnetic field having a flux density of 0·96 T. The conductor forms part of a closed circuit having a total resistance of 0·35 Ω. Calculate:

(a) the e.m.f. induced in the conductor,

(b) the force on the conductor in newtons.

Most relevant sections: **22.5, 17.5,** and **22.4**

E.M.E.U.

14. (a) Sketch the magnetic field due to an electrical current flowing in a solenoid. Indicate the direction of the current and mark the magnetic polarity of the solenoid.

(b) The electrical resistance of a certain solenoid was measured by using an ammeter and a voltmeter connected as shown in Fig. 22.25. When the ammeter reading was 1·5 A the voltmeter reading was 12 V. Calculate

(i) the resistance of the solenoid, and

(ii) the power taken by the solenoid.

Most relevant sections: **22.3, 17.6.1,** and **21.3**

U.E.I.

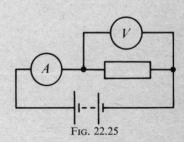

Fɪɢ. 22.25

15. (a) What is meant by a permanent magnet?
(b) Figure 22.26 shows two magnets *A* and *B* placed close together. Sketch the magnetic field that would result (i) if the polarity were as shown and (ii) if magnet *B* were reversed end for end.
(c) Describe, with the aid of clearly labelled diagrams, the construction and principle of operation of any engineering device in which permanent magnets are employed.

Most relevant sections: **22.1, 22.2,** and **22.6**

U.E.I.

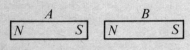

Fig. 22.26

16. (a) State the main difference between a permanent magnet and an electromagnet.
(b) Name two magnetic materials and two non-magnetic materials.
(c) Describe, with the aid of clearly labelled diagrams, the construction and principle of operation of any device in which an electromagnet is used.

Most relevant sections: **22.1** and **22.6**

U.E.I.

17. A rectangular coil wound with 50 turns of wire has a mean width of 15 mm and a mean axial length of 20 mm. The coil is situated in a uniform magnetic field of flux density 0·2 T, with its axis at right-angles to the field. If a current of 10 mA flows through the coil, calculate:

(a) the force acting on one side of the coil, and
(b) the maximum torque on the coil.

Most relevant sections: **22.4** and **9.7**

U.E.I.

18. (a) Sketch the magnetic field due to a current flowing in a solenoid. Indicate the direction of the current and mark the magnetic polarity of the solenoid.
(b) A piece of copper wire is bent into the form of a rectangular loop, mounted on a spindle, and placed between a pair of magnetic poles as shown in Fig. 22.27. The loop has an axial length of 40 mm and a width of 30 mm. If the magnetic field has a flux density of 0·5 T, determine the torque on the loop when a current of 6 A is passed through it.

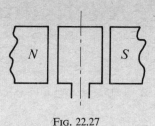

Fig. 22.27

Most relevant sections: **22.3, 22.4,** and **9.7**

U.E.I.

19. By means of a neat diagram and brief notes explain the principle of operation of any *two* of the following:
(a) an electric bell,
(b) a moving coil or moving-iron measuring instrument,
(c) a magnetic chuck incorporating permanent magnets,
(d) an electromagnet of the type which may be used for lifting scrap iron.

Most relevant section: **22.6**

U.E.I.

20. (a) Name two magnetic materials and two non-magnetic materials.
(b) By means of neat diagrams show the type of magnetic field produced either

(i) when an electric current flows through a straight conductor, or
(ii) when an electric current flows through a cylindrical coil of wire.

For each case it is necessary to show the direction of the electric current and the magnetic field.

(c) By means of a neat diagram and brief note explain the principle of operation of *one* of the following:

(i) the magnetic base of a dial test indicator,
(ii) an electric bell,
(iii) a moving-coil or moving-iron measuring instrument.

Most relevant sections: **22.1, 22.3,** and **22.6**

U.E.I.

ANSWERS TO EXERCISES 22

7. 15 N. **8.** 0·5 T. **10. (a)** 0·312 5 N on each side. **(b)** 0·009 375 N m, $\pi/2$ rad. **11.** 0·5 V. **12.** 1·68 V, 8·4 N. **13. (a)** 1·728 V, **(b)** 1·42 N. **14. (b)** 8 Ω, 18 W. **17. (a)** 2×10^{-3} N, **(b)** 30×10^{-6} N m. **18. (b)** 36×10^{-4} N m.

23 Alternating voltage and current

23.1. GENERATION OF ALTERNATING E.M.F.

It has been seen that if a conductor is rotated in a magnetic field of flux an e.m.f. is induced in the conductor and that the magnitude of the e.m.f. will vary instantaneously as the position of the conductor changes. In the single-loop alternator, or generator, the conductor is formed into a rectangular loop and is driven to rotate at constant speed in a uniform magnetic field between a pair of poles as shown in Fig. 23.1. The magnitude and direction of the induced e.m.f. is the same in each side of the loop, and the current taken by the load is collected by carbon brushes (small blocks of carbon) which bear on two brass slip-rings, one connected to each end of the loop with which they rotate.

Consider the loop making one revolution in an anti-clockwise direction as shown in Fig. 23.2. At the instant when the coil is in the position PQ the sides of the loop are moving parallel to the magnetic field; no flux is being cut, and no e.m.f. is induced in the coil. As the loop turns through the first quarter-revolution the magnitude of the e.m.f. increases ($E \propto \sin \theta$; see Chapter 22) until the maximum or peak value is reached when the sides of the conductor are travelling at right-angles to the field, i.e. when the coil has turned through $\pi/2$ rad or 90°. In the second quarter-revolution the e.m.f. diminishes until the sides of the loop are again moving parallel to the field (i.e. until the loop has turned through π rad or 180°) when no e.m.f. is induced in the conductor. The direction of the induced e.m.f. in the first half-revolution is as shown in coil position $P'Q'$ and may be verified by applying Fleming's right-hand (generator) rule. During the second half-revolution the direction of the e.m.f. is reversed in the loop as shown in coil position $P''Q''$, while the magnitude of the e.m.f. varies with the angle turned through as previously described.

The shape of the curve produced is called the waveform of the alternating e.m.f., and the graph in Fig. 23.2 is a sine wave. The equation of a sine curve is

$$y = \sin x,$$

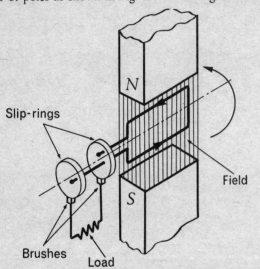

FIG. 23.1. Single-loop alternator

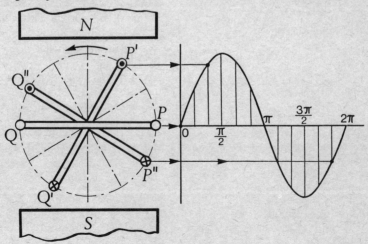

FIG. 23.2. Sinusoidal waveform of induced e.m.f.

so that the magnitude of the instantaneous e.m.f. *e* is given by

$$e = E_m \sin \theta,$$

where E_m = the maximum or peak value of the induced e.m.f.

But, the angle θ turned through in *t* s:

$$= \text{angular velocity } \omega \times \text{time } t,$$
$$\theta = \omega t.$$

Then, $e = E_m \sin \omega t$.

Likewise, the instantaneous current *i*, is given by

$$i = I_m \sin \omega t,$$

where I_m = maximum or peak value of the current.

If the conductor is rotated a number of times, the wave is repeated as shown in Fig. 23.3, and each repetition is called a *cycle*. The time taken for the e.m.f. to change through 1 cycle is termed the *period*, and the number of cycles in each second is known as the *frequency*.

$$\text{Frequency} = \frac{1}{\text{period}}.$$

The unit of frequency is the *hertz* (Hz).

For each revolution of the loop it turns through 2π (rad), and if rotating at an angular velocity ω (rad/s) the frequency *f* is $\omega/2\pi$ (Hz).

Thus the expression $e = E_m \sin \omega t$ may be rewritten as

$$e = E_m \sin 2\pi f t.$$

Also, $i = I_m \sin 2\pi f t$.

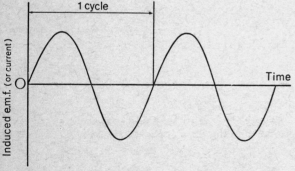

FIG. 23.3. Repetition of waveform

EXAMPLE 23.1. Two sinusoidal currents *A* and *B* have frequencies of 1 Hz and 2 Hz respectively. The peak value of *B* is half that of *A*. Sketch the waveforms on the same diagram. What is the periodic time of *B*?

U.E.I.

SOLUTION

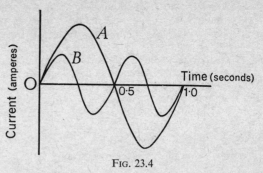

FIG. 23.4

(*Note:* The two waveforms are shown together in Fig. 23.4.)

$$\text{Periodic time of } B = \frac{1}{\text{frequency}}$$
$$= \tfrac{1}{2} \text{ s} = 0 \cdot 5 \text{ s}.$$

EXAMPLE 23.2. An alternating current is represented by the equation $i = 100 \sin 314t$. Determine the frequency and the instantaneous current after 4 ms and 12 ms.

SOLUTION

From the expression $i = I_m \sin 2\pi f t$
$$= 100 \sin 314t,$$
then $2\pi f = 314,$
$$f = \frac{314}{2\pi} = 50.$$

Frequency = 50 Hz.
When $t = 4$ (ms) = $0 \cdot 004$ (s),
$$314 t = 314 \times 0 \cdot 004 = 1 \cdot 256 \text{ (rad)} = 72°.$$
$$\therefore i = 100 \sin 72°,$$
$$= 100 \times 0 \cdot 951\ 1$$
$$= 95 \cdot 11.$$

Instantaneous current = $95 \cdot 11$ A.
When $t = 12$ ms = $0 \cdot 012$ s,
$$314t = 314 \times 0 \cdot 012 = 3 \cdot 768 \text{ (rad)} = 216°.$$

In the third quadrant the sine is negative, and $216° - 180° = 36°$.
$$\therefore i = -100 \sin 36°$$
$$= -100 \times 0 \cdot 587\ 8$$
$$= -58 \cdot 78.$$

Instantaneous current = $-58 \cdot 78$ A.

(*Note:* The negative sign implies that the current is flowing in the reverse direction, i.e. in the second half-cycle.)

23.2. FREQUENCY OF ALTERNATORS

It has been shown that the induced e.m.f. in the single-loop alternator goes through one cycle for every revolution of the coil. Thus for a frequency of 50 Hz, the coil must rotate at a speed of 50 revolutions per second, or 3000 revolutions per minute. In practice, it is undesirable to drive very large alternators at such a high speed, and so multi-polar alternators are used.

For each pair of poles there is 1 cycle per revolution, and for p pairs of poles there are p cycles per revolution. If an alternator has p pole-pairs and if the speed of rotation is n revolutions per second,

frequency = number of cycles per second (Hz)

$$= \frac{\text{number of revolutions}}{\text{per second}} \times \begin{array}{c}\text{number of}\\ \text{cycles per}\\ \text{revolution}\end{array}$$

$$f = n \times p.$$

Throughout the world there are various power supply frequencies employed. For example, the standard frequency in the U.K. is 50 Hz while in the U.S.A. it is 60 Hz.

For a given voltage, a motor designed to run on a 50-Hz system would operate on a 60-Hz supply, but the power would be reduced by approximately 15 per cent, and the speed would be increased by 20 per cent.

EXAMPLE 23.3. An eight-pole alternator is driven at 900 rev/min. Calculate the frequency of the generated e.m.f.

SOLUTION

$$f = \frac{Np}{60} \quad (N = \text{speed in rev/min})$$

$$= \frac{900 \times 8/2}{60}$$

$$= 60.$$

Frequency = 60 Hz.

23.3. STEADY OR R.M.S. VALUE OF AN ALTERNATING QUANTITY

When a direct current of I (A) flows through a resistance of R (Ω) at a potential difference of V (V), the power dissipated is given by:

$$P = I^2R = \frac{V^2}{R} \text{ (W)}.$$

It is seen that for a given resistance, $P \propto I^2$ and V^2. Thus in measuring an alternating quantity it is necessary to obtain that current or voltage which expends the same power as a direct current or voltage in a given resistance.

Consider an alternating current having the waveform shown in Fig. 23.5. If n equally spaced mid-ordinates

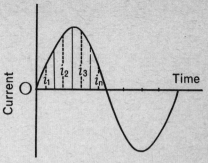

FIG. 23.5. Determination of r.m.s. value

are taken over one half-cycle, the 'steady' or 'effective' value of the current is given by:

$$I = \sqrt{\left(\frac{i_1{}^2 + i_2{}^2 + \ldots + i_n{}^2}{n}\right)},$$

which is the square-root of the mean of the squares of the instantaneous currents. Hence the steady or effective value of an alternating quantity is known as the *root-mean-square* (or r.m.s.) value.

It will be understood that the larger the number of ordinates used, the more accurate the result. The method is of particular advantage for non-sinusoidal quantities.

For a sinusoidal waveform, steady or r.m.s. value = 0·707 × maximum value:

$$E = 0.707 \, E_{\text{m}},$$

$$\text{also } I = 0.707 \, I_{\text{m}}.$$

EXAMPLE 23.4. An alternating e.m.f. with a sinusoidal waveform reaches a maximum value of 325 V during a cycle and has a frequency of 50 Hz. Calculate the periodic time of the cycle and the instantaneous e.m.f. after 1·25 ms. What is the r.m.s. value of the e.m.f.?

SOLUTION

$$\tau = \frac{1}{f}$$

$$= \frac{1}{50} = 0.02.$$

Periodic time = 0·02 s.

The e.m.f. has a sinusoidal waveform so that the equation of the e.m.f. is $e = E_{\text{m}} \sin 2\pi ft$.

$$\text{Then } e = 325 \sin (2\pi \times 50 \times 0.001\ 25)$$

$$= 325 \sin \frac{\pi}{8}.$$

But $\dfrac{\pi}{8}$ rad = $22\frac{1}{2}°$.

$$\therefore e = 325 \sin 22° \ 30'$$

$$= 325 \times 0.382\ 7$$

$$= 125.$$

Instantaneous e.m.f. = 125 V.

For a sinusoidal e.m.f. the steady or r.m.s. value is given by

$$E = 0.707\, E_m$$
$$= 0.707 \times 325$$
$$= 230.$$

R.M.S. value = 230 V.

SUMMARY

The time taken for an alternating quantity to change through 1 cycle is termed the period.

$$\text{Frequency} = \frac{1}{\text{period}}\ (\text{Hz}).$$

For a multi-polar alternator, $f = np$, where p = number of pole-pairs and n = speed (rev/s).

Instantaneous current $i = I_m \sin 2\pi ft$, and

instantaneous e.m.f. $e = E_m \sin 2\pi ft$, where

I_m and E_m are the maximum or peak values of the current and e.m.f. respectively.

The r.m.s. value of an alternating current is measured in terms of the direct current that produces the same heating effect in a given resistance.

By the mid-ordinate method:

$$I = \sqrt{\left(\frac{i_1{}^2 + i_2{}^2 + \ldots + i_n{}^2}{n}\right)}.$$

For sinusoidal quantities,

r.m.s. value = $0.707 \times$ maximum value

i.e. $I = 0.707\, I_m$.

EXERCISES 23

1. Explain the terms cycle, period, and frequency, as applied to an alternating quantity.

Most relevant section: 23.1

2. What is meant by the instantaneous value and the peak value of an alternating current?

Most relevant section: 23.1

3. A conductor rotates at an angular velocity of 157 rad/s between a pair of magnetic poles. Calculate the frequency of the generated e.m.f.

Most relevant section: 23.1

4. What is meant by the r.m.s. value of an alternating current? Why is this value used?

Most relevant section: 23.3

5. A sinusoidal alternating current has a maximum value of 5 A. What is the r.m.s. value of the current?

Most relevant section: 23.3

6. An alternating voltage waveform consists of isosceles triangles and has a peak value of 100 V. Draw the waveform for one half-cycle and estimate the r.m.s. value of the voltage.

Most relevant section: 23.3

7. Sketch the waveform for a sinusoidal alternating current for a period of 2 cycles. If the time for the current to rise from zero to peak value is 5 ms, what is the frequency?

Most relevant section: 23.1

U.E.I.

8. What is meant by the waveform of an alternating current? Draw a half wave for a current with a peak value 5 A and a frequency 50 Hz. What will be the value of current after 4 ms from zero?

Most relevant section: 23.1

U.L.C.I.

9. What is electromagnetic induction and under what conditions does it take place? Describe, with the aid of sketches, how it is used to generate an alternating current.

Most relevant section: 23.1

C.G.L.I.

10. (a) Explain how an alternating e.m.f. may be produced by rotating a coil in a uniform magnetic field. (b) A coil rotates at 50 rev/s in a uniform magnetic field between two poles. State the factors which determine the e.m.f. produced. What is the frequency and the periodic time?

Most relevant sections: 23.1 and 23.2

E.M.E.U.

11. Fig. 23.6 shows a rectangular coil of wire consisting of 200 turns which is rotated at 25 rev/s in a magnetic field of uniform flux density of 0.05 T, the direction of the field being at right-angles to the axis of rotation *XX*.

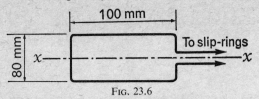

FIG. 23.6

(a) Explain why the e.m.f. picked up at the slip-rings is alternating, and make a neat sketch of its waveform.
(b) Calculate the frequency, and the maximum value of the generated e.m.f.

Most relevant sections: 23.1 and 22.5

C.G.L.I.

12. A coil of 20 turns rotates at 10 rev/s with its axis perpendicular to a uniform magnetic field between two poles of flux density 0·12 T. The active length of the coil side cutting the flux is 50 mm at radius 30 mm.
Determine (a) the maximum e.m.f. generated in the coil and (b) the e.m.f. when the coil is 30° from the position of maximum e.m.f.
Express the generated e.m.f. in the form:
$e = E_m \sin 2\pi ft$, where f = frequency, t = time.

Most relevant sections: 22.5 and 23.1

Y.C.F.E.

13. An alternating e.m.f. with sinusoidal waveform reaches a maximum value of 320 volts during a cycle and has a frequency of 50 Hz. Calculate the periodic time of the cycle and give an expression for determining the instantaneous e.m.f.
What will be the instantaneous value after 2·5 ms?

Most relevant section: 23.1

U.L.C.I.

14. The following table gives instantaneous values of a sinusoidal alternating current over a half-cycle.

Current i (A)	0	38·3	70·7	92·4	100	92·4	70·7	38·3	0
Time t (ms)	0	1·25	2·50	3·75	5·00	6·25	7·50	8·75	10·00

(a) Using suitable scales, draw the waveform of the current for one complete cycle.
(b) By referring to the waveform, show what is meant by the terms 'period', 'frequency', and 'peak value'. State the value of each of these quantities.
(c) Express the current in the form $i = I_m \sin 2\pi ft$.

Most relevant section: 23.1

U.E.I.

15. Describe, with the aid of a series of sketches, how an alternating e.m.f. is produced by rotating a rectangular coil in a magnetic field. Show clearly the directions of the magnetic field, the direction of rotation of the coil, and the direction of the e.m.f. in the coil sides.
Also indicate the coil positions which give maximum and zero e.m.f. respectively.

Most relevant section: 23.1

U.E.I.

ANSWERS TO EXERCISES 23

3. 25 Hz. 5. 3·535 A. 6. 57·7 V. 7. 50 Hz. 8. 4·76 A.
10. 50 Hz, 20 ms. 11. 25 Hz, 12·57 V. 12. (a) 0·453 V,
(b) 0·393 V, $e = 0.453 \sin 20\pi t$. 13. 20 ms, 226·2 V.
14. (b) 20 ms, 50 Hz, 100 A, (c) $i = 100 \sin 314\,t$.

24 Ammeters and voltmeters

24.1. ELECTRICAL MEASURING INSTRUMENTS

An important application of magnetic fields of flux is in electrical measuring instruments, e.g. ammeters and voltmeters. Any such instrument *consists fundamentally* of the following parts:

(1) *a deflecting device* which causes the pointer to move across the scale and indicate the magnitude of the quantity being measured;
(2) *a controlling device* which permits the pointer to move by an amount proportional to the magnitude of the quantity being measured, and which returns it to zero position when the instrument is not in use; and,
(3) *a damping device* which reduces the oscillation of the pointer and other moving parts of the instrument to which it is attached.

There are two main classes of ammeters and voltmeters in common use:

(a) the *moving-iron* type, and (b) the *moving-coil* type.

24.2. MOVING-IRON INSTRUMENT

Although there are two basic types of moving-iron instrument, namely, the repulsion type and the attraction type, relatively few of the latter are now manufactured and for this reason no description of them will be given.

The principle of the repulsion-type moving-iron instrument is shown in Fig. 24.1.

When a current flows through the coil of the instrument it magnetizes both fixed and moving irons in direct proportion to the amount of current flowing. Because *like* magnetic poles repel one another, the 'N' and 'S' poles of the fixed iron repel the 'N' and 'S' poles of the moving iron, causing the latter to move clockwise and so turn the spindle which carries the pointer and control springs. The control springs resist the turning effort in proportion to the distance moved by the moving iron and when the resisting force set up in the springs equals the repulsion force between the irons, the pointer comes to rest at a position on the scale which indicates the amount of current flowing. The damping vane brings the pointer to rest quickly and without undue oscillation, through the resistance to movement of the air in the damper chamber.

If an alternating current flows in the coil, the instrument will still read correctly because both irons are always magnetized so that their adjacent poles are *like* and so repel each other.

24.3. MOVING-COIL INSTRUMENT

The principle of the moving-coil instrument is shown in Fig. 24.2.

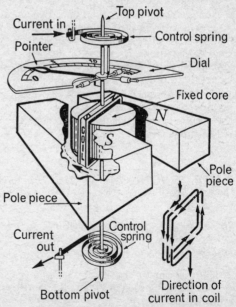

Fig. 24.2. Principle of the moving-coil instrument

It was stated in Chapter 22 that a conductor carrying an electric current when in a magnetic field is subject to a force tending to move the conductor at right-angles to the magnetic field.

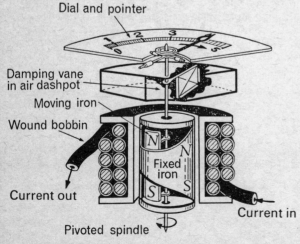

Fig. 24.1. Principle of the moving-iron instrument

In the moving-coil instrument the current to be measured passes via the control springs through the moving-coil, two sides of which are in the field provided by a permanent magnet. Uniformity of flux is ensured by the fixed core around which the coil moves. The motion of the coil in the magnetic gap is resisted by the control springs proportionately to the angle of deflection, and the coil and pointer come to rest when the resisting force set up in the springs equals the deflecting force on the sides of the coil. Eddy currents induced in the aluminium former when moving in the magnetic gap set up a force tending to oppose the motion of the coil, thus providing a *damped* movement.

The direction of movement of the coil depends on the direction of the current flow relative to the direction of the magnetic field. As the latter is constant in direction, reversal of direction of the current flow will cause the coil to move in the opposite direction. For this reason a permanent-magnet moving-coil instrument can be used to measure direct current only.

24.4. PIVOTLESS INSTRUMENTS

In a pivoted instrument there are several components that can fail in service: and sometimes do. The pressure on the pivot is something like 1·5 GPa; spring-loaded jewels help to protect the pivots but even so they can become mushroomed or chipped, especially when they are subjected to shock or much vibration. Even in a new and perfect pivot bearing there is inherent friction which can only increase with age. Control springs can acquire undesirable deformations.

In pivotless instruments there are no pivots, and hence no pivot-friction, no jewels, and no control springs. A beryllium–copper ribbon provides both suspension and, by its twisting, the control torque.

Fig. 24.3 shows the principle of the moving system of a pivotless instrument.

24.5. COMPARISONS BETWEEN MOVING-IRON AND MOVING-COIL INSTRUMENTS

The main advantages and disadvantages of the moving-iron instrument are as follows.

Advantages:

(1) suitable for use in d.c. circuits or a.c. circuits;
(2) more robust construction than the moving-coil instrument; and
(3) cheaper than the moving-coil instrument.

Disadvantages:

(1) non-uniform scale; and
(2) more easily affected by *stray* (external) magnetic fields than the moving-coil instrument.

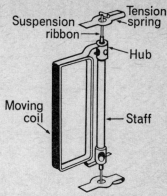

FIG. 24.3. Moving system of a pivotless instrument

The main advantages and disadvantages of the moving-coil instrument are as follows.

Advantages:

(1) uniform scale;
(2) well shielded from stray magnetic fields; and
(3) highly sensitive.

Disadvantages:

(1) suitable for use in d.c. circuits only; and
(2) more expensive than the moving-iron instrument.

24.6. EXTENSION OF INSTRUMENT RANGES

Basically, a given instrument can be used as an ammeter or a voltmeter, and can be adapted to operate over more than one scale range.

24.6.1. Ammeters. A suitable resistor of *low resistance*, known as a *shunt*, is connected in *parallel* with the instrument so that the latter carries only a small proportion of the current to be measured. It will be clear from Fig. 24.4 that a current I flowing in a circuit will divide at junction A, I_g passing through the instrument, and I_s through the shunt. These currents then re-unite at junction B.

$$I = I_g + I_s \quad \text{and} \quad V = V_g = V_s.$$

24.6.2. Voltmeters. By connecting a suitable resistor of *high resistance*, termed a *multiplier*, in *series* with the

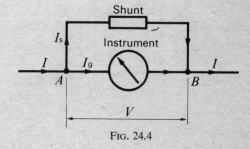

FIG. 24.4

instrument the latter will have across it a small proportion of the total voltage. Fig. 24.5 shows that the sum of the p.d.s across the instrument and multiplier equals the p.d. between points C and E, i.e. the p.d. to be measured.

$$V = V_g + V_m \quad \text{and} \quad I = I_g = I_m.$$

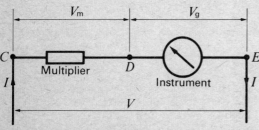

FIG. 24.5

EXAMPLE 24.1. A galvanometer has the following specifications: resistance of coil $= 3\ \Omega$; current required for full scale deflection $= 25$ mA. It is desired to use the galvanometer (a) to measure a maximum current of 5 A and (b) to measure a maximum voltage of 25 V.

In each case show how this may be done and state the value of any additional resistance used.

U.L.C.I.

SOLUTION

(*Note:* The diagrams are shown in Figs 24.4 and 24.5.)

(a) For full-scale deflection,

$$V = I_g R_g$$
$$= 0.025 \times 3 \text{ V},$$

p.d. across instrument $= 0.075$ V.

The shunt is in parallel with the instrument,

$$\therefore \text{ p.d. across shunt} = \text{p.d. across instrument}$$
$$= 0.075 \text{ V}.$$

$$\text{Current through shunt } I_s = \text{total current } I - \text{current through instrument } I_g$$

$$I_s = 5 - 0.025$$

$$\text{Current through shunt} = 4.975 \text{ A}.$$

$$R_s = V/I_s$$

$$= \frac{0.075}{4.975} = 0.015\ 08.$$

$$\text{Resistance of shunt} = 0.015\ 08\ \Omega.$$

(b) The multiplier is in series with the instrument,
∴ current through multiplier

$$= \text{current through instrument}$$
$$= 0.025 \text{ A}.$$

$$\text{P.D. across multiplier } V_m = \text{total p.d. } V - \text{p.d. across instrument } V_g$$

$$V_m = 25 - 0.075$$
$$= 24.925.$$

$$\text{P.D. across multiplier} = 24.925 \text{ V}.$$

$$R_m = \frac{V_m}{I_m}$$

(where R_m = resistance of multiplier)

$$= \frac{24.925}{0.025} = 997.$$

$$\text{Resistance of multiplier} = 997\ \Omega.$$

SUMMARY

Ammeters and voltmeters consist essentially of the following parts: (1) a deflecting device, (2) a controlling device, and (3) a damping device.

The two main classes of ammeters and voltmeters are: (a) the moving-iron type and (b) the moving-coil type. The range of an ammeter scale is extended by connecting a shunt (low resistance) in parallel with the instrument. The range of a voltmeter scale is extended by connecting a multiplier (high resistance) in series with the instrument.

EXERCISES 24

1. Give one advantage and one disadvantage of the moving-iron type of electrical instrument.

Most relevant section: 24.5

U.E.I.

2. State briefly why a moving-coil instrument is not suitable for measurements with alternating current.

Most relevant section: 24.3

U.E.I.

3. State which type of ammeter, moving-iron or moving-coil, may be used to measure (a) a.c. and (b) d.c.

Most relevant section: 24.5

U.L.C.I.

4. What is the function of a multiplier resistance for use with a voltmeter? Draw a diagram to illustrate its use.

Most relevant section: 24.6.2

U.L.C.I.

5. What is the function of an ammeter shunt? Draw a diagram showing an ammeter connected to a shunt.

Most relevant section: 24.6.1

U.L.C.I.

6. With the aid of sketches, describe the construction of a moving-iron repulsion-type meter. Show clearly the method used to produce controlling torque and damping torque.

In an instrument of the above type the coil resistance is $25\,\Omega$ and requires $30\,mA$ for full-scale deflection. Determine the value of the series resistor required for the instrument to be used over the range 0–20 V.

Most relevant sections: 24.2 and 24.6.2

U.E.I.

7. (a) Explain, with the aid of a sketch, the principle of operation of a moving-coil ammeter.
(b) An ammeter of resistance $1\cdot5\,\Omega$ has a range 0–2A, and a resistor of $0\cdot5\,\Omega$ is used with it for the purpose of increasing its range. Show, by a diagram, how the instrument and resistor should be connected in a circuit. Having made this connection, the ammeter now reads $1\cdot2\,A$. What is the current in the circuit?

Most relevant sections: 24.3 and 24.6.1

C.G.L.I.

8. A moving-coil instrument gives full-scale deflection when the current through it is 15 mA and the p.d. across it is 6 V.

(a) Calculate the resistance of the instrument.
(b) Find the value of a resistor required to enable the instrument to read 24 V at full-scale deflection and show by a sketch how it should be connected.
(c) Find the value of a resistor required to enable the instrument to read $1\cdot0\,A$ at full-scale deflection and show by a sketch how it should be connected.

Most relevant sections: 17.5, 24.6.1, and 24.6.2

C.G.L.I.

9. A moving-coil instrument has a resistance of $5\,\Omega$ and requires a potential of 75 mV to cause a full-scale deflection.
(a) Calculate the value of an additional resistance to enable the instrument to read: (i) up to 1 A; and (ii) up to 30 V.
(b) Draw circuit diagrams showing how the resistances would be connected.

Most relevant sections: 17.5, 24.6.1, and 24.6.2

N.C.T.E.C.

10. The coil of a moving-coil instrument has a resistance of $20\,\Omega$ and is fully deflected when a current of $2\cdot0\,mA$ flows through it. Calculate the values of resistors required to enable the instrument to be used: (a) as a 0–0·5 A ammeter, and (b) as a 0–10 V voltmeter.
In each case show by a diagram how the resistor would be connected.

Most relevant sections: 17.5, 24.6.1, and 24.6.2

Y.C.F.E.

11. (a) A fully loaded 500-V d.c. motor has an output of $18\cdot65\,kW$ and an efficiency of $74\cdot6$ per cent. The input to the motor is measured by an ammeter and a voltmeter, both instruments being of the moving-coil type and fully deflected at the stated load. The moving coil of each instrument has a resistance of $5\,\Omega$ and each gives full-scale deflection when its terminal p.d. is $0\cdot075\,V$. Calculate the resistance of the additional resistor in use with each instrument.
(b) In each case show by a diagram how the resistor is connected to the instrument.

Most relevant sections: 21.3, 24.6.1, and 24.6.2

U.E.I.

12. Two identical moving-coil instruments have the following specification:

Resistance of moving coil = $5\,\Omega$.

Current required for full-scale deflection = 15 mA.
It is desired to use one instrument to measure currents up to 50 A and the other instrument to measure voltages up to 500 V. In each case show, by means of a circuit diagram, how this may be done and calculate the resistance of the additional component employed.

Most relevant sections: 24.6.1 and 24.6.2

U.E.I.

13. (a) What is the function of an ammeter shunt? Draw a diagram showing how a shunt should be connected to an ammeter.

(b) The resistance of a coil is measured by the ammeter-voltmeter method. With the voltmeter connected across the coil, the readings on the ammeter and voltmeter are 0·5 A and 4·0 V respectively. Calculate the approximate resistance of the coil. If the resistance of the voltmeter is 500 Ω, find the true resistance of the coil.

Most relevant sections: 24.6.1 and 17.6.1

U.E.I.

14. (a) What is the function of an ammeter shunt? Draw a diagram showing how a shunt should be connected to an ammeter.

(b) The resistance of a certain resistor is unknown. Using B.S. symbols, draw a circuit diagram to show how the following apparatus should be connected to an electrical supply in order that the approximate resistance of the resistor may be obtained:

  (i) an ammeter,
  (ii) a voltmeter,
  (iii) a variable resistance,
  (iv) the resistor of unknown value.

What is the purpose of the variable resistor? How is the approximate resistance of the unknown resistor calculated from the readings of the instruments?

Most relevant sections: 24.6.1 and 17.6.1

U.E.I.

ANSWERS TO EXERCISES 24

6. 641·7 Ω. **7.** 4·8 A. **8. (a)** 400 Ω, **(b)** 1200 Ω. **(c)** 6·08 Ω. **9. (a) (i)** 0·076 Ω, **(ii)** 1995 Ω. **10. (a)** 0·080 34 Ω, **(b)** 4980 Ω. **11.** 0·001 501 Ω, 33 328 Ω. **12.** 0·001 501 Ω, 33 328 Ω. **13. (b)** 8 Ω, 8·13 Ω.

25 Related chemistry

25.1. CHEMISTRY

Chemistry is a branch of science which is concerned with the study of the structure, composition, and properties of substances and the changes they undergo. The relationship between chemistry and engineering is particularly significant since engineering is also concerned, in part, with the properties and behaviour of materials. Familiar examples which illustrate this relationship are the extractions of metals from their ores, corrosion, and the combustion of gases.

25.2. MATTER

Matter is the term used to describe any naturally occurring substance. Physically, matter is found in one or more of the following forms—solid, liquid, or gaseous. These three possible states of existance may be defined as follows:

(a) solids have fixed shape and volume;
(b) liquids have fixed volume but their shape depends upon the form of the vessel in which they are contained; and
(c) gases have neither specific shape or volume for they will completely fill the containing vessel.

There is a considerable degree of interchangeability between these three states of matter, and the factors which determine in what form a substance will exist are those of temperature and pressure. Water, for example, at low temperatures becomes a solid (ice); whilst at higher temperatures it will become steam (a vapour). Metals, normally solid, can be liquefied by raising their temperatures to above their melting point. Gases may be liquefied by appropriate changes in temperature and pressure.

25.2.1. Elements. Many naturally occurring substances, such as iron ore, water, or salt, are not 'pure' substances and can be broken down into a simpler form of matter. For example, iron ore is mined in several forms, usually as an impure oxide of iron, i.e. a chemical combination of iron and oxygen; salt is a chemical combination of sodium and chlorine; and water consists of hydrogen and oxygen chemically combined. These substances, hydrogen, oxygen, iron, and chlorine are called elements. An element is a substance which cannot be decomposed into any simpler substance by chemical change. There are 105 known elements, of which about 90 occur naturally in the earth and its atmosphere. All other substances are formed from a combination of two or more of these elements.

25.2.2. Atoms and molecules. All elements are made up of atoms each element having its own particular form of atom. An *atom* is the smallest part of an element which can take part in a chemical change. It consists of a central nucleus around which particles travel in orbit. The nucleus contains positive charges of electricity known as *protons* together with neutral charges of electricity known as *neutrons*. The particles in orbit are negative charges of electricity and are given the name electron. *Electrons* have a mass of approximately 9.11×10^{-31} kg whilst the mass of a neutron and a proton are each nearly 1850 times greater than the mass of the electron. It can be seen therefore that practically the whole mass of the atom is concentrated in the nucleus.

The simplest structure is that of the hydrogen atom which consists of one proton in the nucleus and one electron in orbit. It is the only atom which has no neutron in the nucleus. Next to hydrogen comes helium which has two protons and two neutrons in the nucleus and two electrons in orbit. In all atoms the number of protons on the nucleus is equal to the number of orbiting electrons making the atom electrically neutral.

The *atomic number* of an element is the number of protons in its nucleus, and in a neutral atom is also equal to the number of electrons. The *relative atomic mass* of an element is the average mass of its atoms on a scale in which one atom of the ^{12}C isotope of carbon has a mass of 12 units. ^{12}C isotope of carbon has six protons and six neutrons in the nucleus.

Isotopes are forms of the same element in which the atoms have different masses because they contain different numbers of neutrons in the nucleus.

Today heavy hydrogen and heavy water are quite well known expressions. As previously stated hydrogen itself has no neutron in its nucleus, but heavy hydrogen or deuterium has one neutron in its nucleus. Heavy water is made up of heavy hydrogen and oxygen and is present in ordinary water to the extent of about 1 part in 6000 parts by mass.

Most elements have natural isotopes, but the separation of these is very difficult because their chemical properties are unchanged and their physical properties are very little different.

25.2.3. Symbols. A chemical symbol represents one atom of an element, and whenever possible the initial letter of the element is used. In certain instances, to avoid confusion, the first two letters of the name are used or the Latin equivalent. Some of the more common elements

together with their chemical symbol and approximate relative atomic mass are given in Table 25.1.

TABLE 25.1

Element	Symbol	Relative atomic mass
Aluminium	Al	27
Carbon	C	12
Chromium	Cr	52
Cobalt	Co	59
Copper	Cu	63·5
Gold	Au	197
Hydrogen	H	1
Iron	Fe	56
Lead	Pb	207
Magnesium	Mg	24
Manganese	Mn	55
Mercury	Hg	200·5
Molybdenum	Mo	42
Nickel	Ni	59
Nitrogen	N	14
Oxygen	O	16
Phosphorus	P	31
Silicon	Si	28
Silver	Ag	108
Sulphur	S	32
Tin	Sn	119
Tungsten	W	184
Vanadium	V	51
Zinc	Zn	65

A *molecule* is the smallest part of an element or compound (see below) which can exist in a free state whilst retaining the characteristic properties of the substance.

25.2.4. Compounds. A compound is formed when two or more elements combine chemically. The chemical combination takes place during a chemical reaction in which the atoms of one element combine with atoms of one or more other elements. The molecule of a compound will contain at least one atom from each of the two or more elements from which the compound is formed. A typical simple compound is water (H_2O), in which one atom of oxygen combines with two atoms of hydrogen to form one molecule of water. This combination is shown diagrammatically in Fig. 25.1.

Note: The atom cannot exist in a free state and can only take part in a chemical reaction as part of a molecule.

Valency is the term used to denote the relative combining powers of different atoms. It is usual to express the combining powers in terms of hydrogen atoms. The valency of an element is the number of atoms of hydrogen which combine with, or are displaced by, one atom of the given element.

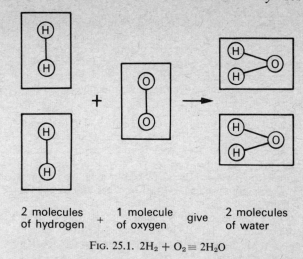

| 2 molecules of hydrogen | + | 1 molecule of oxygen | give | 2 molecules of water |

FIG. 25.1. $2H_2 + O_2 \equiv 2H_2O$

Calculations in chemistry often involve the use of relative molecular masses of both elements and compounds.

The *relative molecular mass* of an element is equal to the relative atomic mass multiplied by the number of atoms per molecule. The relative molecular mass of a compound is the sum of the relative atomic masses present in the compound.

The chemical symbols and approximate relative molecular masses of a number of elements and compounds are given in Table 25.2.

TABLE 25.2

Element or compound	Symbol	Relative molecular mass
Acetylene	C_2H_2	26
Carbon	C	12
Carbon monoxide	CO	28
Hydrogen	H_2	2
Oxygen	O_2	32
Sulphur	S	32
Sulphur dioxide	SO_2	64
Water	H_2O	18

25.2.5. Mixtures. The ratio in which masses of various elements combine to form compounds is fixed by natural laws and is in no sense a haphazard affair. For example, iron combines with sulphur in the ratio of 4 to 7 by mass. This means that 4 kg of iron will just combine with 7 kg of sulphur. Any attempt to combine these two elements in a different ratio would result in a surplus of one of the elements after the reaction had taken place. There are, however, some substances which will not combine chemically with others and therefore are termed mixtures.

A mixture contains two or more different substances (either elements or compounds) which are not chemically combined. Examples of mixtures are air which is a mixture of oxygen, nitrogen, and argon together with traces of other gases; soluble cutting oil emulsions which are mixtures of oil and water; and iron filings and sawdust.

25.2.6. Alloys. If some pieces of zinc are added to a quantity of molten copper the zinc will be dissolved and a solution obtained. On cooling the mixture solidifies and the solid substance formed is termed an alloy. In this case the alloy is brass. Generally an alloy is formed if two or more metals are fused together and allowed to solidify.

Other important alloys include bronze (an alloy of copper and tin) and the 'white' metals, used for bearings etc., in which various amounts of tin are dissolved in molten lead.

25.3. PHYSICAL AND CHEMICAL CHANGES

In a physical change the molecules of all the substances involved remain unchanged, i.e. no new substance is produced.

An example of a physical change is the familiar workshop 'suds'. A soluble cutting oil is added to water to form a mixture or emulsion, the white and soapy appearance of this emulsion being totally different in appearance to its constituents, oil and water. However different the emulsion may look, the oil and water in it still retain their respective properties. These are the water to cool the cutting operations and the oil to lubricate the passage of the metal chips over the cutting tool face. The mixing of the oil and water to form an emulsion has produced a physical change but no chemical reaction has taken place.

In a chemical change the molecules of the new substance are different to those of the original substances. A new substance is formed having new and different properties to those of the original substances. The rusting of iron illustrates a chemical change in which iron and oxygen react together under certain circumstances to form the familiar red oxide of iron commonly called rust.

25.4. COMPOSITION OF AIR

Air is a complex mixture of several different gases. Its approximate composition by volume is

nitrogen	78 per cent
oxygen	21 per cent
argon	0·9 per cent,

the remaining volume consisting of traces of carbon dioxide, hydrogen, neon, helium, krypton, and xenon. In addition, there are relatively small proportions of other gases, including sulphur dioxide, together with water vapour in varying proportions. It is because air has slight variations in its composition that it is identified as a mixture and not a compound.

In industrial areas, particularly, there are minute quantities of carbon monoxide and sulphur dioxide in the atmosphere. Carbon monoxide is a poisonous gas which is formed by the incomplete combustion of petrol in motor car engines and, of course, if inhaled in any quantity, is injurious to health. Sulphur dioxide is an undesirable acid gas which is partly converted into sulphuric acid by rain-water and plays a significant part in the corrosion of metals and attacks stonework.

The presence of water vapour in the air is variable but is of great importance. Man's physical comfort is dependent upon the relative humidity of the air, which is a measure of its water content. On 'muggy' days air is almost saturated with water vapour and evaporation from the human skin becomes very slow; this creates discomfort and a certain degree of tiredness.

Again dust in the atmosphere apart from being a health hazard, presents a direct problem to the engineer. Considerable damage can be done to surfaces sliding over one another, particularly in the case of bearings. Dust can block oilways and so prevent efficient lubrication or cooling. In addition it is extremely undesirable where accurate measurements are required. Modern air-conditioning plants are designed to remove dust from the air and to control temperature and moisture.

25.5. OXIDATION AND REDUCTION

To the chemist the idea of oxidation and reduction are far more complex than the descriptions which follow would suggest. However, for ease of understanding and application to common engineering situations the following definitions should suffice.

Oxidation refers to a chemical reaction in which oxygen is added to an element or compound.

Reduction is in effect the reverse of oxidation being a chemical process in which oxygen is removed from a compound.

Both oxidation and reduction play important roles in engineering processes. Most metals combine with oxygen to form oxides. For example, aluminium remains unaffected in dry air; in moist air a superficial skin of oxide forms which protects the metal from any further action under normal conditions. One very important application of reduction is the extraction of a metal from its ore. Zinc, iron, and lead compounds are more readily reduced than other metallic compounds so it is economical to produce these metals by using a comparatively cheap reducing agent, coke.

A further example of oxidation occurs in the heat-treatment of metals. An important factor in this treatment is the furnace atmosphere, i.e. the type of gas present in the heat-treatment furnace during the process. If too much oxygen is present in the furnace this will combine with the metal components to form metallic oxides which will appear as a scale. This effect is most undesirable and particularly in the case of steel not only does the scale need removing, but the carbon content of the outer layers will have been reduced, and this can seriously affect the properties of the steel. Under these conditions the atmosphere is said to be oxidizing. When it is necessary to avoid oxidation in a heat-treatment process the furnace should have a reducing atmosphere or contain a neutral atmosphere by use of an inert gas.

25.6. CORROSION OF METALS

Corrosion may be defined as the chemical reaction between a metal and its environment as a result of which the metal becomes partially converted to a chemical compound such as an oxide.

In the process of this conversion metals lose their essential qualities of strength, ductility, and elasticity.

The necessity to restrict or prevent corrosion where possible can be seen by the fact that hundreds of millions of pounds are lost annually in the U.K. owing to the wastage of materials and the failure of expensive equipment through corrosive action. Fortunately, corrosion is a surface chemical reaction only and there are available a number of ways of slowing it down or stopping it completely.

25.6.1. Rusting of iron. The rusting of iron is one illustration of the corrosion of metals. If four separate quantities of iron filings are left for a week in dry air, moist air, air-free water, and tap water respectively, it will be found that the filings subjected to the moist air and tap water have rusted, while the other two samples have remained rust free. From this experiment it will be observed that for the rusting of iron to occur both air or oxygen and water must be present.

Tne chemical composition of rust is that of a hydrated oxide of iron, i.e. iron oxide chemically combined with some molecules of water. The oxide film has a porous structure and is only loosely attached to the iron beneath so that the rusting process continues unchecked.

Corrosion of metals is a complex subject, but it is known to be electrolytic in nature and is caused by the natural formation of *voltaic or corrosion cells* on the surface of the metal. Such cells require an electrolyte which in general is readily obtainable from the atmosphere in the form of moisture together with sodium chloride (common salt), carbon dioxide, and sulphur dioxide.

25.6.2. Electrochemical series. It is an experimental fact that when a piece of metal is placed in water an e.m.f. known as the electrode potential is set up and acts at right-angles to the surface of the metal in contact with the water. The magnitude of this electrode potential is dependent upon the metal and the purity of the water, etc. In Table 25.3 a selection of metals taken from the electrochemical series is chosen with the e.m.f. expressed in volts.

TABLE 25.3

Metal	Normal electrode potential (V)
Silver	+0·8
Copper	+0·34
Hydrogen	0·00
Lead	−0·13
Tin	−0·14
Zinc	−0·76
Aluminium	−1·67
Magnesium	−2·34

The tendency of a metal to corrode is indicated by its position in the electrochemical series. In general the metals at the positive end of the table are more resistant and those at the negative end are more likely to corrode. It must be emphasized, however, that the position of a metal in the table is only an approximate guide to the relative tendency of the metal to corrode as other factors must be taken into account. For example, some metals form protective oxide layers which can cause the metal to be much more resistant to corrosion than the table would suggest.

25.6.3. Zinc as a protective coating. Zinc is widely used for metal-coating purposes or *galvanizing* as it is called. The zinc coating is usually applied by hot dipping or electrolysis.

If the surface of the galvanized iron is scratched so as to expose the iron then because of possible moisture and acid gases in the atmosphere an electrolytic cell is set up—zinc–acid–iron. In this case the zinc passes into solution and not the iron. Rusting of the iron is therefore prevented at the expense of the zinc. In this situation the zinc is described as a sacrificial metal.

The disadvantages of zinc-coatings are the rather uneven and darkening of zinc surfaces with the passage of time and the fact that zinc is readily attacked by acids and alkalis.

25.6.4. Tin as a protective coating. Tin has many advantages as a protective surface since it is extremely tough, adheres extremely well to base metals, and withstands extensive deformation. It is also very resistant to most corrosive agents and, because its salts are non-toxic, tin is widely used in the food industry. Its main disadvantage is that if the surface of tin-plated iron becomes scratched, corrosion of the iron is very rapid. The reason for this is that in the presence of an electrolyte, such as acidified moisture from the atmosphere, an electrolytic cell is set up—tin–acid–iron. In this case the iron passes into solution, i.e. the iron rusts.

Note the opposite effects of zinc and tin coatings when scratched and their relative positions in Table 25.3.

25.6.5. Corrosion of aluminium. Aluminium is very much more resistant to corrosion than steel because of the formation of a thin coating of oxide. However, the corrosion of aluminium is markedly increased when it is in contact with other metals. If, for example, aluminium is used in conjunction with ferrous metals in the form of brackets, couplings, and bolts, the aluminium sometimes tends to corrode at the point of contact. In the presence of carbonic acid (water plus carbon dioxide) or sulphurous acid (water plus sulphur dioxide) both of which are normal contents of an industrial atmosphere, an electrolytic cell (aluminium–electrolyte–iron) is formed. In this situation it is found that the corrosion of the aluminium is accelerated and the rusting of the iron is retarded. In a similar manner, if aluminium is coupled to copper or a copper alloy, very rapid corrosion takes place at the point of contact due to the formation of corrosion cells.

Anodizing is a process by which aluminium can be made more resistant to corrosive action. It involves the formation of a relatively thick and coherent oxide film on the surface of the aluminium by making the metal the anode of an electrolytic cell and passing a current. The anodized coating is then sealed by subjecting it to heat in the form of hot water or steam. This treatment increases the strength of the oxide film and improves its resistance to corrosion.

25.6.6. The prevention of corrosion. A number of methods are used industrially to protect steel from corrosion some of which are given below:

(1) painting;
(2) galvanizing—zinc-coating;
(3) tin-plating;
(4) calorizing (aluminium-coating);
(5) sherardizing (zinc-dust-coating);
(6) nickel- and chromium-coatings;
(7) prevention of corrosion cells by preventing contact between different metals.

Further information on the above processes will be found in specialist textbooks on materials and corrosion.

25.7. COMBUSTION

Combustion, or burning, occurs when a substance combines with oxygen accompanied by the evolution of heat and light. Before combustion takes place it is generally necessary for intense local heat to be applied to raise the substance to its ignition temperature.

Carbon and hydrogen can combine in a variety of proportions and in many different ways. The compounds formed from these elements are grouped together under the heading *hydrocarbons*. Most fuels belong to this group including petroleum and coal. The combustion of these and associated fuels provide the heat energy which on conversion to mechanical energy furnish the main energy requirements of industry.

Elements will combine only in certain ratios. When hydrocarbons burn, therefore, a precise amount of oxygen is required to combine with the carbon and hydrogen. If there is too much oxygen present the products of combustion will be carbon dioxide (CO_2), water vapour (H_2O), and a surplus of oxygen. When insufficient oxygen is available, for complete combustion the final products could contain carbon monoxide (CO) instead of carbon dioxide (CO_2). For most practical purposes, oxygen required for combustion is taken from the atmosphere. In certain applications however, such as oxyacetylene welding, pure oxygen is used. The use of oxygen in this form ensures better combustion and provides a far hotter flame than would be the case if oxygen from the air were used.

SUMMARY

The atomic number of an element is the number of protons in its nucleus.

The relative atomic mass of an element is the average mass of its atoms on a scale in which one atom of the ^{12}C isotope of carbon has a mass of 12 units.

Isotopes are forms of the same element in which the atoms have different masses because they have different numbers of neutrons in the nucleus.

Valency is the term used to denote the relative combining powers of different atoms.

Relative molecular mass of an element = relative atomic mass × number of atoms per molecule.

Relative molecular mass of a compound = sum of relative atomic masses present in the compounds.

Physical change—molecules of all the substances involved remain unchanged.

Chemical change—molecules of the new substances are different to those of the original substances.

Oxidation—a chemical reaction in which oxygen is added to an element or compound.

Reduction—the reverse of oxidation in which oxygen is removed from a compound.

Corrosion of metals is electrolytic in nature. It may be defined as the chemical reaction between a metal and its environment.

Combustion or burning occurs when a substance combines with oxygen accompanied by the evolution of heat and light.

EXERCISES 25

1. (a) What are the three forms in which matter can exist? Give one engineering application of each.
(b) Name two substances which can exist in more than one of these forms. Upon what factors does the form of the substance depend?

Most relevant section: **25.2**

2. (a) Illustrate, by use of a labelled diagram, the structure of a hydrogen atom.
(b) Explain the essential difference between an atom and a molecule.
(c) What is meant by an isotope? Give two examples of isotopes.

Most relevant sections: **16.2** and **25.2.2**

3. Define the following terms: element, compound, mixture, and alloy. Give one workshop example of each.

Most relevant sections: **25.2.1, 25.2.4, 25.2.5,** and **25.2.6**

4. Classify the following under the headings elements, compounds, and mixtures:

Air	Acetylene	Steel
Aluminium	Carbon dioxide	Sulphur dioxide
Brass	Iron	Water
Brine	Rust	White metals
Bronze	Soluble cutting oil	Zinc

Most relevant sections: **25.2.1, 25.2.4,** and **25.2.5**

5. Distinguish clearly between a physical and a chemical change. Describe an example of each which may occur in the workshop.

Most relevant section: **25.3**

6. What is meant by the terms oxidation and reduction? Give engineering applications of each.

Most relevant section: **25.5**

7. (a) Name four metallic elements.
(b) Name four non-metallic elements.
(c) Name four compounds, stating the elements of which each is composed.
(d) Give two examples of a mixture.
(e) Give two examples of oxidation.

Most relevant sections: **25.2.1, 25.2.4, 25.2.5,** and **25.5**

U.E.I.

8. (a) Distinguish between chemical change and physical change. Name two examples of each type of change.
(b) Name two corrosive gases commonly found in the air in industrial areas, and describe (i) their effect on metals, (ii) one method of preventing this effect.

Most relevant sections: **25.3, 25.4,** and **25.6.1**

U.L.C.I.

9. (a) What is corrosion?
(b) What is rust? Your answer need not refer to chemical symbols but it should state the essential ingredients to be present for the formation of rust.
(c) How can atmospheric conditions affect the rate at which steel parts tend to rust?
(d) Will rust form on a steel bar placed in: (i) dry air, (ii) air-free water?
(e) Name a metal which does not rust.
(f) Give three examples of protective coatings used to prevent or minimise corrosion.
Credit will be given for reference to appropriate laboratory experiments.

Most relevant sections: **25.6, 25.6.1, 25.6.3,** and **25.6.4**

E.M.E.U.

10. Sheet iron rusts in use less rapidly when galvanized than when tinned.
Comment on the significance of this observation.

Most relevant sections: **25.6.3** and **25.6.4**

11. (a) Define the term 'combustion'.
(b) What are the main combustible elements likely to be found in a fuel?
(c) Why is sulphur one undesirable element in a fuel?

Most relevant sections: **25.7** and **25.4**

12. (a) Name the main constituents of air and the percentages, by volume, in which they are present.
(b) Explain fully why dust particles are undesirable in the atmosphere making particular reference to engineering situations.

Most relevant section: 25.4

13. Explain why aluminium appears to have a good resistance to corrosion. Briefly indicate a process by which this resistance is improved.

Most relevant section: 25.6.5

14. Select one method of corrosion protection mentioned on p. 184 and write a report on the process chosen for a particular engineering situation. Indicate in the report the main advantages of the method chosen and why it is considered suitable for the project involved. Reference should be made to a number of relevant sources of information.

Most relevant section: 25.6.6

15. Explain what is meant by the term 'atmospheric pollution'. What has been the effects of the introduction of 'smokeless zones' in industrial areas?

Most relevant sections: 25.4 and 25.6

INDEX